Broilers For Profit
From The Experiences of The Pioneer Broiler Chicken Raisers of This Country

by Michael K. Boyer

with an introduction by Jackson Chambers

This work contains material that was originally published in 1897.

This publication is within the Public Domain.

*This edition is reprinted for educational purposes
and in accordance with all applicable Federal Laws.*

Introduction Copyright 2017 by Jackson Chambers

Self Reliance Books

Get more historic titles on animal and stock breeding, gardening and old fashioned skills by visiting us at:

http://selfreliancebooks.blogspot.com/

Introduction

I am pleased to present yet another title on Poultry.

The work is in the Public Domain and is re-printed here in accordance with Federal Laws.

As with all reprinted books of this age that are intended to perfectly reproduce the original edition, considerable pains and effort had to be undertaken to correct fading and sometimes outright damage to existing proofs of this title. At times, this task is quite monumental, requiring an almost total "rebuilding" of some pages from digital proofs of multiple copies. Despite this, imperfections still sometimes exist in the final proof and may detract from the visual appearance of the text.

I hope you enjoy reading this book as much as I enjoyed making it available to readers again.

Jackson Chambers

PREFACE.

The writer of this book has always maintained that broiler raising is a profitable branch of poultry culture, and should, to a certain extent, be part of the work upon every poultry farm; but, as an exclusive affair, there is more or less risk.

The following pages have been written and compiled not only from the author's personal experience, but also the experiences of the pioneer broiler raisers of the country.

Hammonton, New Jersey, practically laid the foundation for the success of this branch of poultry work. It had a hard road to travel—inasmuch as it had it all to learn by dear experience. Those who will criticise the efforts and failures of Hammonton do not stop to thank her for what she has taught the poultry world. Men have failed here by the scores; but every failure taught a lesson to some one. To show just what Hammonton did, and the valuable lessons she has taught, this book has been written.

Respectfully,

MICHAEL K. BOYER.

Hammonton, N. J., Jan. 1, 1897.

CONTENTS.

CHAPTER I.
The Business as It Is — Why Men Fail at It — The Capital, Land, and Time Required to Make It a Success 5

CHAPTER II.
The Value of Incubators — Points of the Different Machines — Hints in Running Them — Mistakes as Made by Amateurs 9

CHAPTER III.
Disappointing Hatches — Brooders and Brooding System 13

CHAPTER IV.
Different Methods of Feeding — Some of the Ailments — Hints on the Breeding Stock 18

CHAPTER V.
More About Incubators — The Question of Moisture — Answers to Queries . . 22

CHAPTER VI.
Dressing and Shipping to Market — Several Valuable Queries 27

CHAPTER VII.
More Questions and Answers — Henry Nicolai's Broiler Plant — A German's Profitable Combination. 30

CHAPTER VIII.
An Interview With P. H. Jacobs — Questions Answered 33

CHAPTER IX.
The Model Brooder House — Hammonton Criticised — Experiences — Fattening Broilers — Questions and Answers 40

CHAPTER X.
Mr. Jacobs and His Opinions — Richard H. White's Methods 44

CHAPTER XI.
The Visitors — Albert Reed's Broiler and Egg Farm — Combinations — Colonizing Broilers 49

CHAPTER XII.
Geo. W. Pressey's Plan for Raising Broilers — Harry Philips' Gigantic Establishment — Valuable Pointers — Gov. Levi P. Morton's Plan — Broilers and Ducklings — How to Get Fertile Eggs 53

Notes in Passing 60

BROILERS FOR PROFIT.

CHAPTER I.

THE BUSINESS AS IT IS.—WHY MEN FAIL AT IT.—THE CAPITAL, LAND AND TIME REQUIRED TO MAKE IT A SUCCESS.

Broiler raising, like all other industries, is one that requires both skill and capital. No person can hope to be successful without either. No business on God's earth, for that matter, will flourish without skilled labor and money back of it. The worst blow that this industry ever got was the big boom sent out some years ago by men with "axes to grind." It was reported to be the business for the masses. The easy work, the great demand, the "millions in it," all tended to draw in raw recruits. Men came to Hammonton with scarcely enough cash to build a decent brooding house, fully determined to "do or die,"—and they died! Where one succeeded half a dozen failed.

A little common sense should teach anyone that the business that stood gaping for operators, that would flourish with the care of every Tom, Dick or Harry—that was "a gold mine" to all who touched it, certainly would soon be overdone, and in a very short time the supply would be double the demand; but as it is, the demand cannot be reached by one-third. The repeated failures keep up the prices, and those who have the ability to "hold on" are the people who are making it pay.

The way the broiler business is generally run counts much against the profits. It has been said that it pays better to buy the eggs than raise them. The way the broiler men of Hammonton are scouring the country "for eggs," willing to pay more than the market price, and then unable to buy the quantity needed, shows that that is a sad mistake. Broiler men have been known to spend a whole day looking for eggs, and return with hardly enough to half fill one incubator.

Chicks hatched from these boughten eggs, as a rule, come out in "all styles and shapes." Some are of quick growing, others of slow growing breeds. Some are from healthy parentage, others from poorly conditioned fowls. Fifty per cent of fertile eggs has become a good test. Fifty per cent of market stock has become the average success.

Does not that argue in favor of home raised eggs? Would not the percentage be a third larger if the proper stock, feed, and care were employed? Eggs can be raised at a cost of one cent each. They bring two cents apiece in the market for table use. Incubator men pay a few cents more per dozen.

All over the country this broiler fever has gathered in men of wealth. Some have undertaken it for pleasure, some for profit, some for both! To run their establishments men have been secured at good salaries, and every convenience and appliance secured — and yet the cry is, "we cannot meet expenses!" Why is it? We could assign several reasons: In the first place, these men employed are very often not thorough. They probably have had success in a small way, but to undertake the management of a large concern places them outside their territory. Again, these men are often careless. They know how to manage the business, but they have not the ability "to get there," to use a vulgarism. Another class of men do not mind the "small things," because the boss is rich and can stand the small losses.

We know of an instance where a young man was employed to manage the brooding house of a rich gentlemen who cared principally for his own table supply, yet that object could not be secured. While the employe in question had sufficient knowledge of the business, he had a bad habit of continually putting off work, thinking there was plenty of time yet. Result, the gentleman had to *buy* broilers for his own table—and the young man lost his job!

We mention these instances to show that the only way to make a success of *any* branch of the poultry business, the work must be closely inspected or done by the owner; and the more experience he has, the better will be the results.

Starting broiler farms on rented grounds is another sad mistake, and one which is only equaled by going into debt. Both are millstones. Renting a farm makes an expense that must be met regularly, and one which makes deep inroads on the profits. The man who has not enough cash to pay for his plant, and run it without an income for at least six months, had better keep his hands off.

Summing up the business — looking at both the bright and dark sides — we find there is a good living ready for the man who has pluck, energy and cash. There must be personal work put into it, and all unnecessary expenses kept down.

It is amusing to read the estimates given by men who think they know all about the expense of raising broilers up to marketable size. One party puts the cost as low as six cents per pound, many at ten cents, and few at fifteen cents. Carefully interviewing the different broiler raisers of Hammonton, we find that it cannot be done at less than fifteen cents a pound — thirty cents for a two pound bird. By getting down to close figuring we find that on an average, nineteen and a half cents profit for *each bird* is about all that can be realized. In this estimate we voice the sentiments of every broiler raiser we have visited, with one exception — J. C. Browning, — who says he has kept a careful account, and finds that his average is twenty cents. Place that estimate alongside the figures given in boom articles! It is one thing to build air-castles, and another to get at practical results.

The broiler business could have no more staunch friend than the writer; but to misrepresent matters, and draw into service men and women unfit for the work or not properly equipped, is not only unjust to them but cripples the business. That is one of the main reasons why people fail at it.

As an adjunct to other branches of the poultry business, or to fruit growing or trucking, broiler raising is profitable. In fact, it makes a splendid crop, because the expense of living, etc., is not entirely placed upon it. The man who attends to them is not compelled to ask the business to pay his entire family expense — and many more items are evenly divided, which give the profits a better chance. Henry Nicolai, of Hammonton, has a broiler plant of one thousand capacity. He runs it from December to July; and last year cleared four hundred dollars in his establishment. During the summer he raises truck. The profits of both give him a comfortable living. Henry Phillips raises truck and fruit in summer, and broilers in winter. Albert Reed farms in summer—and so we might go through the whole list. It goes to show that as an *exclusive* business it does not pay.

Is it not asking too much of any branch of farming to expect sufficient profit to keep a man in idleness six months of the year? We remember reading some years ago, that " during the summer months the broiler men do nothing, having made enough during the winter to keep them comfortable the entire heated term," or words to that effect. The truth is, that there is not a single instance in Hammonton where such a comfort is enjoyed. We have been making close investigations, and the general verdict is: " It is a good winter job," and so it is. More cannot be expected.

" How much money do I need to start in the broiler business?" is a query that comes to us about as regular as our taxes. It reminds us of a reply we received when we were in our teens, and suffering with the orange culture fever. Having read glowing accounts of the immense profits in raising oranges, and the little labor attached to the business, we wrote to a friend then residing in Florida, (and, who several years before that left home " to become rich " by the industry), asking " how much money was required?" His reply was: " I came here with a thousand dollars, spent all that, and could spend five thousand more." It cured our fever.

So it is with broiler raising. To start in big, a commencement can be made with a thousand dollars—but you may need five thousand more. Our advice has always been, start small, and grow up with the business. Have you three hundred dollars, and an income besides? Use one hundred to build a small brooding house; one hundred for incubators and other necessary fixtures; one hundred for working capital, and your outside income for your living. Save all the returns from the business that year, and enlarge the next. The profits and experience from five hundred broilers during the first year, will fit you for raising a thousand the second year. In a few years you will have established a sufficiently large farm to enable you to devote your time during the winters; but hold on to some other summer employment. If you have five hundred dollars, invest three hundred as aforesaid, and one hundred in good laying fowls, and another hundred in suitable buildings — and each year your farm will substantially grow.

It is a hard matter to set down just how much money a man should have. Some men can do more with one hundred dollars than others can with twice that amount. It is not so much the amount of money you put in the business, as how you place it, and the knowledge you have to properly run the machinery after you have it all in readiness.

"How much land do I need?" That is more easily answered. A building lot will hold a good sized brooding house and runs, but a building lot will not make a successful broiler farm. The man who must depend upon the "luck" he has in buying up eggs, generally has a hard road to travel. He may be fortunate enough to have some one raise good eggs for him, but the profits he will have to pay on those eggs draws from the profits he would have on his broilers. We have known forty cents per dozen to be paid for eggs in winter. Any good poultryman can raise them for twelve cents per dozen. The egg man makes more on his eggs than the broiler man makes on his broilers. Again, we have seen sixty per cent of boughten eggs cast aside after being tested in the incubator, owing to their being either perfectly clear or containing a weak germ. From the forty per cent left, we have known but two-thirds to bring forth good strong chicks. Is there any profit in that? Think of forty cents per dozen for eggs, and only thirty per cent hatch.

A good poultryman would know how to mate, feed and care for sufficient hens to get fertile eggs for his incubators, and produce strong healthy chicks. Instead of a building lot for a broiler farm, we say by all means a five acre tract. Divide four acres into sixteen quarter acre each yards, and in each keep twenty-four two year old hens and three vigorous cockerels to every two pens, a plan which we explain in Chapter III. — three hundred and eighty-four hens in all, and you can supply a large brooding house. Quarter-acre yards could be added, and a small flock of birds would be unable to kill the grass. Or, it could be sown to rye, and all winter they would have necessary green food.

Five acres are just the right number. Even if a man did not care to keep more than two hundred fowls, he would have two acres for fowls, one acre for residence and brooding-house, and two acres to raise potatoes, cabbages, turnips and other necessary winter food for both old and young stock.

Of course a farm this size would keep him busy all the year round, but he would be on the safe side. Out of broiler season he could sell eggs and surplus old fowls, and thus swell the receipts. In short, we cannot see how a broiler farm can be much of a success without an egg farm attached. We fully understand that this presentment of facts will not coincide with what has so repeatedly been published; but if our reputation is to be attacked by our determination to tell the truth, we are ready for the battle. The views given are not only as we personally have found them, but as our neighbors — Hammonton's well advertised broiler raisers — also give them. Broiler men have nothing to advertise. They have no object in giving the business an unhealthy boom. They are satisfied with their *job* — but they are not growing rich.

Broiler raising has not made a Vanderbilt, a Gould, or a Sage; but it has made honest farmers. Coupled with an egg farm they have had good profits.

They live comfortably; their diet is not composed of all sauces and puddings; they delight in their brown bread and beans. They dress not in the height of fashion, but, honest like, they wear plain but good clothing. Their complexions are of a healthy color, their constitutions are strong, their hearts are happy because their work is their love, and the air-castles are "busted." They no longer dream of a fortune in broilers, but they sleep the sleep of the just, because they now know that hard work wins, and they can get a good day's wages for every day they are at work.

CHAPTER II.

The Value of Incubators.—Points of the Different Machines.—Hints in Running Them.—Mistakes as Made by Amateurs.

"Are incubators a success?" writes a correspondent. Yes — but they are not always successfully run. So often an enthusiast branches out in the papers that his hens beat the incubators, seeming to think that the idea of the invention of incubators was to beat the record of the hens — in other words, to bring out more chicks, and out of such eggs that the hens are unable to hatch. Such a thing is impossible. No machine was ever invented, nor ever will be, that can do better work than a faithful broody hen.

Incubators have a different mission. Before their invention it was next to impossible to get out broilers in time to meet the long prices. To have sitting hens in mid-winter is quite a trick. Even a few sitting hens in mid-winter would not do much towards filling a small brooding house; but the wholesale work of the incubators tells. Chicks not by the dozen, but by the thousands! At almost any time from December to June, a thousand or more eggs are at one time in course of incubation in a single broiler house in Hammonton.

An objection was once raised against artificial incubation in the belief that the chicks were not as vigorous as those brought out by hens. Every year we personally use both methods. During the winter and early spring we run our incubators, and after the hens become broody we set them, and slack up on the machines. The chicks in the brooders are kept separate from those raised by the hens, and we have never yet been able to notice a particle of difference. Why should there be? The incubator is constructed after the fashion of the hen. It hatches the eggs at the same temperature as does the hen. If the heat has been regular, and the amount of moisture and ventilation sufficient, the chicks are sure to come out as strong and rugged as when hatched by a hen. There can be no difference.

In no way can the usefulness of the incubator be belittled. It has been a god-send to the poultry world, and the only agent by which the raising of broilers can be profitably conducted.

There are many machines upon the market. Some we know to be excellent, some good, while some are but ordinary, if not worthless. It is not the object of the writer to praise any one make, or decry some he has not tried. This book will not solve the query, "Which is the best?" Such a conundrum cannot be answered.

Let us make a general review: In the first place, a machine should be well constructed. No matter how good the various parts may be, if the general frame of the incubator is poorly gotten up, the working will be a failure. We one day purchased a machine that seemed to be a model in every point of construction, but a few days running warped the top, and the hatch was spoiled. The lumber used was not thoroughly seasoned. We mention this more as a pointer for the manufacturer than the novice.

There is no doubt about it, every year there are machines put upon the market that should never have existed. They are built to sell, and not to successfully work. If the writer is able to throw out hints that will steer amateurs away from bogus incubators, his aim will be fulfilled.

Mr. Rankin once said, in referring to the great distrust of incubators by poultrymen: "This is not strange, for there are so many worthless machines upon the market which hatch well on paper, but whose only mission seems to be to disappoint their purchasers and addle his eggs, that it has made people skeptical. I do not wish a hen around in any other capacity than that of an egg producer."

In the town of Hammonton, N. J., there has been a variety of experience. Almost every make has been tried here. Some houses use the Hammonton hot water incubators; some the Hammonton hot air; some the Prairie State; while the Eureka, Pineland, Monitor, Keystone, Excelsior and Paragon have their friends.

Having a reliable machine, the next thing necessary is to properly manage it. The manufacturer's directions are not always plain. Sometimes they are wrong. Shun such testimonials as "a child can run it." Make up your mind that it takes brains to run even the best automatic machines in the world. Do not believe that any machine can hatch every fertile egg. There never was a hen that could do that. How can we expect an invention by man to do what nature seems to have been unable to perform? Do not be led away by the assertions that the machines require but little care each day. Broiler men are continually watching their incubators. Here in Hammonton even the very machines that are advertised to be next to perfect as self-regulators, must be continually looked after. If they are neglected the broiler men know disaster may follow. It is as Geo. W. Pressey, of this place, says: "The best regulated machine in the country needs regulating."

Writers, and all who ever used incubators, agree that a dry house cellar is the proper place to run the machines. Yet this is not always the proper thing to do, on account of the objection raised by insurance companies. They will assume no risks on a house in which a machine is run; but we can see no danger. The average incubator lamp is more carefully attended to, and in a safer condition for burning all night than the ordinary house lamp, yet people with sickness or children never hesitate to let the lamp burn all night, and the insurance company raises no objection. Where it can possibly be done, an outside cellar should be built. A regular cellar can be dug in the ground and walled up with stones Frame sides two or three feet high can be erected upon the stone wall, and a roof over it, and you have the best incubator room that could be secured. Where this is not possible, a dry room without any fire in it will do very well. Quite a number of the broiler men of Hammonton run their machines in the room where the large stove is that heats the brooders. The result is they must continually be on the watch, and considerable of the time thus employed could be avoided if the aforesaid cellar were used. Machines must never be set where there can be draughts; and care must be taken that they stand perfectly level.

Where to place the thermometer, is another important point. After we have made our second test, (on the fourteenth day), and removed all the infertile eggs, we place bulb of the thermometer on a live (fertile) egg. It is a mistake to place the thermometer otherwise. We put all the fertile eggs in the center of the machine (having removed the dead ones) and place the thermometer in the center of these. That plan is the safest to pursue.

Moisture is an unsolved problem. That it is necessary there is not the least bit of doubt, but just how much is needed, cannot be set down in print. It must be determined by surrounding circumstances. An incubator placed in a cellar would naturally require less moisture than one in a room in which fire is kept. Again, some incubators need more moisture than others. In Hammonton moisture is not applied to the Hammonton machines until after the fourteenth day. In some of the automatic machines it is applied a little at the beginning, and gradually increased. Some machines have the bottom of the egg chamber covered with sand, and this is kept wet, or rather damp, the entire hatch. The exact quantity of moisture necessary is best determined by testing the eggs. The air cell on the fifth day should measure about a quarter of an inch; on the tenth day, a half inch; on the fifteenth day about five-eighths of an inch; and about three-quarters of an inch on the nineteenth day — the measurement taken in the middle of the egg. Such air cells indicate the proper amount of moisture. If less than that, too much moisture is given; if more, there is a lack of moisture.

The eggs must be turned after the fourth day. This is a point upon which all incubator men agree. Some machines use roller trays to turn them; others turn the eggs with a crank; and others supply an extra tray which they lay on the tray of eggs to be turned, catch it in the center and turn over. No matter how it is done, the eggs must be turned every morning and evening, about twelve hours apart; — but when the chicks begin to pip the shell, they must not be handled.

It is necessary to test the eggs. White shelled eggs can be safely examined on the fourth day, but it will take seven days before a dark shelled one can be satisfactorily tested. We test all eggs on the seventh day for the first, and on the fourteenth day for the second or last test. By this time we know exactly the condition of all the eggs in the incubator, and as all the dead or rotten eggs are removed we have not the stench in the machine so common when the second test is not made. Egg testers are cheap — for that matter every machine has one sent with it. Experts, however, seldom use a tester. They take the egg in the left hand and hold it between forefinger and thumb, the former on the big end and the latter under the small end of the egg. The right hand then shades the light, and the egg is looked at through the sides. To test in this way the room must be dark excepting what comes from the lamp used to test with. The perfectly clean eggs that are tested out on the seventh day are not spoiled, but can be safely used in the kitchen. The broiler men sell them to the bakers, who use them in making pastry. The eggs tested out on the fourteenth day we feed to our fowls. We beat up a few every other day in the soft food.

It is a hard matter for beginners to understand that hatches must not be tampered with. When machines with glass doors in front are used there is not so much danger, as the amateur can pretty well content himself by looking through the glass; but in machines using drawers, the curiosity of the operator too often causes him to open and see how the hatch is progressing. It is needless to say that these draughts of cold air are damaging to the hatch. Keep the machine closed as much as possible. Either operate it yourself, or have one person attend to it. It never pays to have two persons attending the same machine, as something is sure to be neglected. Always keep the doors of the drawers closed when the eggs are out being examined, so as to keep up the temperature. It is easier to reduce the heat in the egg chamber than to increase it. When the eggs are hatching is the time the machine must be kept closed. Let the youngsters kick about all they want to. As long as the chicks do not gasp for breath they will not be injured by the heat. We let them remain in the egg chamber until thoroughly dried off. Then they are removed to a nursery. They must not be taken out too fast, as every time chicks are removed the temperature in the chamber is reduced.

The same rule that applies to the poultry business in general applies to the broiler business—*begin small*. Herein is where the amateur makes a grave mistake. He is too ambitious. He dreams of machines that hold a thousand eggs, and brooding houses that hold five thousand chicks, etc. If they have the money, ten chances to one they will begin large—and then, not having the necessary experience, the concern " busts." We believe in one and two hundred egg machines. In case of accident the risks are not so great. By starting three small machines the bad eggs can be tested out, and all put into two incubators, and the third one re-started.

Lamps must be attended to every day. The crust on the wick should be scraped off every evening, the lamp refilled, and then carefully wiped off so that not a particle of oil or dirt is left to catch fire. Beginners are so apt to overlook this work, resulting in disasters. Experts place the care of the lamp as the most important in the operation of incubators.

In testing eggs the amateur can readily detect the condition. A perfectly clear one is infertile. There is no germ there, and it will not hatch. The good strong eggs are those that show a regular "spider"—a small round speck from which red veins run in all directions looking like the legs of a big spider. The larger this spider the stronger the egg. If the germ dies during the incubation, it will be generally known by a distinct red line in the form of a circle about the size of a silver quarter. As the hatch progresses the eggs become darker. After the fourteenth day, when the veins have entirely disappeared, and the germ is seen to float about when the egg is turned, it shows that it has died, and the egg had better be removed.

At the end of twenty-one days all hens' eggs should be hatched. If the machine was mismanaged, the hatch will necessarily be prolonged, and the chicks weakened. Good strong eggs hatch very often on the nineteenth day. It is a mistake to help a chick out of the egg. If it is too weak to come out of its own accord, it will be too weak to be of any good when it is hatched.

Finally, follow out the directions of the manufacturer. He is supposed to know the weak and the strong points of his machines, and if his incubator is a good one, and he himself a practical poulterer, success will certainly follow.

CHAPTER III.

DISAPPOINTING HATCHES. — BROODERS AND BROODING SYSTEM.

What seems to puzzle the amateur most is to know why he cannot, despite the best of care, secure percentages in hatches anywhere near what is openly done at the shows in incubator displays. A FARM-POULTRY subscriber would like to know how it is that incubator men can secure such grand hatches in the show room, securing percentages that are not equaled by any one else. About the best reply that can be given, is a quotation from a letter written by George W. Pressey, of Hammonton, N. J., to the manufacturer of a well known machine, who challenged Mr. Pressey to an incubator contest. The quotation says :—

"Remember, however, that testing two or three times, and the last time on the nineteenth day, when it is possible to remove every egg that has not a live chicken in it ready to come out the next day or two, is no test of the incubator, but merely tests the tester, and the skill of the operator; and if we let the eggs alone, then we only learn who had the best eggs. To really test an incubator, then, we must know that the eggs were exactly alike in both machines. Is this possible?"

Leaving the subject of incubators for the present, we must turn our attention to brooders and brooding systems. It is much easier to hatch out the chicks than it is to raise them. Especially is this so in midwinter or early spring.

It is if anything more important to have a good reliable brooder than it is a hatcher, and in the selection of a particular kind good judgment should be used. The best brooder is the one that is fashioned the nearest after the hen. Here in Hammonton there has been a variety of experience in this line. Both top and bottom heat are employed, with the preference for the former. At first, single brooders were used — heated by lamp. As this necessitated considerable labor, especially where a dozen or more brooders were in use, the Packard system began to be adopted. This system gives bottom heat, the brooders being heated by hot water pipes running under them. Finally the Smyrna system (top heat) was introduced, which is gradually being adopted by the broiler men here.

If we must pattern after the hen in order to have success in artificial methods, is it not reasonable to suppose that *top heat* is the proper system to employ? Does not the natural mother give warmth to her young from her body while sitting *over* them? It is logic. Yet the advocates of bottom heat have a strong plea; they say, "When you wish to warm yourself, which do you do — put your head or your feet to the fire?" Then comes the top heat man with the warning that bottom heat produces leg weakness; and the bottom heat advocate assures us that top heat causes head troubles. Thus the doctors disagree. But, summing it all up, the top heat men are in the majority, and it will not be long until it will be universally employed.

Nearly all the incubator manufacturers have nurseries for sale. They are invaluable for chicks up to two weeks of age, by which time they are strong and better able to take care of themselves. Then they are removed to the regular brooding houses. Those nurseries that have neither top nor bottom heat, but a free circulation of hot air throughout the entire chamber, are the safest to use, inasmuch as the chicks have a more uniform heat. All nurseries should have a shallow pan of water kept constantly in the chamber. Too dry an air is apt to cause leg troubles.

In the construction of brooding houses there is considerable room left for improvement. In the first place, the manner in which they were formerly made gives entirely too much glass to them. What let in the warmth during the day while the sun was shining, would also admit the cold at night. That trouble, however, was somewhat regulated by having regular window shades which were pulled down under the glass at night. Mr. Pressey used oiled muslin instead of glass, and says he liked it much better, as it more readily kept out the cold than glass did.

The subject of ventilation in the brooding houses is one that has been discussed almost to death. In the debate some good things have been said, and likewise some very foolish ones. Theories on both sides of the question have been used. There must, undoubtedly, be ventilation; but care must be taken that whatever process is used will not allow draughts. Ventilators on top of the roofs are risky. Holes made on the sides of the house up near the peak, is a good method.

In our brooding houses we let the air in principally through the exit doors leading to the outside runs. In most houses these doors are not more than a foot square, and when opened in the morning there is a big rush and squeeze to get out. We have doors a foot high and the width of the brooders. They are made of 2 x 3 inch frame, covered with half-inch mesh, and over this wire we tack white muslin. This plan, aided by the aforesaid holes on the sides of the brooder up near the roof, gives us all the ventilation we need, and there is never that sickening air in the building so common in houses where the life-giving oxygen is kept out.

The wide doors we refer to, leading to the runs, serve another good purpose. When we open them in the morning, we give ample room for every chick to at once run out. There is not the least bit of danger of accident which often occurs when small doors are used. Besides, with a door the width of the brooder thrown open, the entire interior is at once aired and cleared of whatever gases may have arisen.

Accustom chicks to fresh air after two or three weeks of age, and there is less mortality, and stronger stock. We believe that improper brooding has killed more chicks than the incubators have failed to hatch.

When the chicks are hatched they are naturally weak and inexperienced. If a nursery is convenient, they will, for the first two weeks, gain strength very rapidly. Begin the heat at ninety degrees, and keep it as near that as possible for the first week or ten days. Then gradually reduce until (after the chicks are removed to the large brooder) they become accustomed to a temperature of seventy degrees, which should be when about three weeks of age. What a mistake to begin at one hundred, and thus compel the little ones to endure torture instead of comfort. This high temperature is what makes weak and delicate chicks. In some large brooding houses, where nurseries are not used, boards are stood up around the brooder about a foot away. In this way the chicks are gradually accustomed to their foster mother, and cannot wander to the other end of the pen during the night, only to become chilled and die. Chicks must be taught. At first it is necessary to place them under the brooder so that they know what to do to keep warm. As the kindly call of the natural mother is absent in the brooder the keeper must attend personally to the little ones until they know of the inviting comfort their silent parent has for them.

Thermometers can and should be used to determine the heat, but after the chicks are made to "feel at home," a better sign that all is well is the manner in which the chicks act. If, when closing up the house for the night, it will be seen that the little ones are stretched out on their brooder floor, with their bills buried in the sand, we know that nothing more can be done for them; everything is right. If, on the other hand, they crowd up together, unsettled, there is not enough warmth. Or, if they sit with their mouths open, the heat is too great. These signs are indications of affairs, and must be watched closely.

Brooding houses are divided off into pens. They average about 5 x 7 feet each. Into pens like these we have seen as high as one hundred and fifty chicks crowded. Practically, they were built for one hundred chicks, but fifty will do better in them. Overcrowding is as big a mistake in brooding chicks as in housing hens. There is no economy in close quarters. Small flocks always do best.

Cleanliness is a most important rule in brooding. The dirt must be collected daily. Upon this, however, we shall more properly dwell in our general remarks upon the care of chicks.

The houses must be built so as to be proof against both rats and dampness. To do this, they should be set about a foot above ground. From the sills, half-inch wire should be stretched from side to side, and firmly tacked with wire staples. Upon this one inch flooring nailed; the floors of the pens covered with sand several inches thick. Dry road dust is also good. It is a pleasure to see how the little ones enjoy this. They will dust themselves on warm days, and take an endless amount of exercise.

Fresh air is one of the agents that produces good sound broilers. For the first week or ten days chicks thrive very well in close quarters. They can stand considerable heat, and must be kept indoors; but when they grow stronger, which would be when about two weeks old, they need the fresh air. On nice warm days, they must have the run of the little yards outside their pens. The experiment has several times been tried to keep them indoors from the time they were hatched until ready for market, but the theory failed.

Mr. Pressey, on warm days, removed the muslin sash, and threw open his entire house, letting in both air and sunshine.

But the chicks must not be let outdoors before the sun is up, or while there are dews or fogs; nor must the little ones be allowed outside on windy or damp days. Common sense is necessary. They will mind sudden changes quite readily.

Artificial heat in the brooder house, outside of what is given from the brooders themselves, is not generally required. It is better not to have any extra heat. The chicks soon learn their foster mother. They exercise in a temperature several degrees colder, and when uncomfortable run back to the "mother" for extra warmth. With the pens several degrees colder than the brooders there is a good preparation for outside runs. Let the chicks out of hot brooding pens into the much colder temperature of the outside air, and colds and disease are quickly invited.

Speaking of close confinement calls to mind an experiment we tried last year with a few ducklings. Ducklings will stand such experiments better than chicks. From the time they were hatched until three months of age, we kept them indoors. We supplied green food, and the regular bill of fare. Nothing was lacking in that line, but we could not see any progress in growth worth mentioning. They fattened very nicely, but the frame was stunted. At three months of age we prepared for a feast of roast duckling, but we were sadly disappointed. Instead of the sweet firm meat, we had carcasses of fat that gave the meat an oily flavor. The flavor of choice duck meat was gone, and our faith in open air and exercise was more firmly strengthened than ever. We have got the best of nature in many ways by our brooding methods, but we need her fresh air all the same. Those who raise hogs know the difference of meat in those allowed to root and exercise from those kept quiet and fattened.

Brooder raised chicks, if properly cared for, should thrive as well if not better than those raised by hens. If the eggs are washed in lukewarm water before put into the incubator, and if the brooder house is kept scrupulously clean, there cannot be the least chance for lice. Lice slaughter more chicks annually than any other cause. When lice are found on brooder chicks it is the result of neglect. Nothing will breed lice faster than a sitting hen, and as biddy is of no use in artificial methods, there is a chance to keep clear of vermin. Allow only a few lice to begin operations in a dirty brooding house, and disaster will quickly follow.

Never put chicks hatched by hens in a brooding house with those brought out by incubators. One lousy chick will soon spread vermin among the entire flock. A friend of the writer, while living in Virginia, had a fine lot of chicks in his brooder that were growing nicely and in the best of health. One day he secured a few chicks from a neighbor who unfortunately lost the mother hen, and, as he could find no lice upon the little ones, he put them in his brooder with the other chicks. In the course of a few weeks trouble began. One after another of his chicks began dying off, and before he really found out what the trouble was, his entire brooder was full of lice.

With the same precaution regarding lice, we must watch the temperature of the brooders. As we said before, no brooder should start at a temperature above ninety degrees. Several years ago we had a small flock of chicks in a brooder, with which we tried the experiment of high temperature. We began with one hundred degrees. All was well for the first week, then we noticed an unusually rapid growth — heat and good food did it — then a sudden weakness, and before the chicks were three weeks old they lost control of their legs, and squatted about entirely helpless. Since then we have adopted the milder treatment, and have had good success ever since.

CHAPTER IV.

Different Methods of Feeding. — Some of the Ailments. — Hints on the Breeding Stock.

An incubator may hatch ever so well, and a brooder do the finest kind of hovering, yet if the chicks are not properly fed, there certainly cannot be any kind of success. Good feeding tells. There is no fixed bill of fare, and in taking up this subject, we can only give our own experience, and what we have observed on other farms in Hammonton.

As is well known, the chick comes from the shell without the least bit of appetite. Scientists say that nature has given it the power to absorb the contents of the yolk prior to its coming out of the shell. Consequently, it is not necessary to give any food for the first twenty-four hours. Some writers say thirty-six hours after hatching is early enough for the first feed; but we always dish up the first meal after they are a day and a night old. After we hatch out the chicks we put them in a nursery, where we keep them for the first ten days or two weeks. After that they are removed to the regular brooding house. Upon the floor of this nursery we cover about a half inch of bran, so that when the chick is ready to eat it will find some food right before it. Then, in a little trough we place rolled oats, or pin-head oatmeal, and subsequently begin, say about a week afterwards, a mash feed. We might as well add here that we also give stale bread crumbs to alternate with the rolled oats, until they have their stomachs more fortified for heavier food. Some of the broiler men in this town grind up, or finely crack, whole wheat, which they feed instead of the rolled oats. We like both methods; but think more favorably of the oats diet. We also give boiled milk as a drink.

Some writers think that for the first few days it is best not to give any drink. We differ. Deprive the youngsters of something to drink and they will fairly gorge themselves when allowed to get to the water. We do not, however, start with water; we boil milk, and give that instead.

There used to be an opinion that hard boiled eggs was the best food to start with; but we think that too much reliance had been placed upon that diet. It has been proved beyond a doubt that an excessive use of hard boiled eggs will produce bowel troubles. There can be no harm in a judicious use of them; but we have found more virtue in saturating bread crumbs with a fresh egg.

Two parts of bran, and one part of corn meal, scalded several hours before using, is an excellent feed after the chicks are a week old. But two parts bran, one part of corn meal and one part ground wheat, is better. A little of meat scraps — say a handful to a pail of the above mixture—should also be added.

After two weeks of age, cracked wheat and cracked corn makes a substantial meal. From the start, grit of some kind must be within reach. Some use fine flint; some finely cracked oyster shells; and some fine gravel. It matters not what kind is used. Powdered charcoal should also be kept in a little box in the pen so the chicks can help themselves.

Green food, as chopped up onion tops, or cabbage leaves, are very beneficial. Lettuce can be raised early in the season in hot-beds; and a better and more tender plant cannot be found. Where greens are scarce, roast potatoes, cut into halves, furnish a grand substitute; and even when greens are fed, roasted potatoes give an extra treat.

Corn meal, in some form or other, is the "staff" upon which to grow good broilers. It can be used in regular johnny-cakes, or what is known as southern corn bread.

The secret that E. C. Howe, (who ran at one time the largest brooding house in Hammonton), kept until he moved out of the town, was as follows: "One pint corn meal, one teacupful bran, one tablespoonful ground meat. Mix thoroughly. Then take one raw egg, half teaspoonful of baking soda, and one teacup of cold water. Mix these together in a separate dish, and add to the meal, bran and meat. Also put in three tablespoonfuls of ground bone. Bake in a deep pan for two hours. When cool, crumble up for them."

Mr. Howe fed this until the chicks were ten days old, when he gradually introduced a feed of ground wheat, oats and corn, moistened. He always kept ground oyster shell, bone meal and charcoal before them in separate dishes.

To have seen his brooding house, especially in the height of the season, would have been enough recommendation for his manner of feeding.

Jacob J. Peterson, of Vineland, N. J., adopts this plan:

The first day he feeds stale bread crumbs, moistened with milk, alternating with plain stale bread crumbs every two hours. This feed he continues for ten days, after which he gradually withdraws the bread and milk, and feeds ground grain,—one-third corn, and two-thirds wheat, morning, noon and night. No water for the first ten days, but all the skimmed milk they want.

We might add, however, that we never give milk to the chicks until it is boiled. There is less likelihood of having any evil effects from it. We have known of bad cases of bowel troubles coming from feeding fresh milk.

George W. Pressey, of Hammonton, who, with the assistance of his two daughters, raised and marketed nearly five thousand chicks in a single season, used this plan of feeding:

When the chickens are twenty-four hours old, feed them with baked corn cake made as follows: three quarts corn meal, one quart wheat middlings, one cup of meat meal. Mix quite stiff with water or skimmed milk, in which have been mixed four tablespoonfuls of vinegar, and two teaspoonfuls of soda. Bake, and when cold, crumble fine and feed for the first week all they will eat, or during the time they are kept in a warm room, which must never be over ten days, or they will sicken and die for want of pure outdoor air. For the first week they should be fed once a day with mashed potatoes, given plenty of water to drink, and plenty of coarse sand. The feed for outdoors is two parts corn, one part wheat, and one part oats, ground together quite fine. To each ten quart pailful of this mixture add one quart of wheat bran, half a cup of pulverized bone meal, one pint of middlings, and a pint of meat meal. Mix rather dry with hot water, and leave for two hours before feeding, to give it a chance to swell. With this feed, he also, once a week, gives a half teaspoonful of salt, and in cold weather a quarter teaspoonful of red pepper; and once or twice a week he adds a spoonful of sulphur; and about as often, mixes in the drinking water for the day, a spoonful of Douglas mixture to every one hundred chickens. Powdered charcoal is kept before them all the time.

About a week or so before the chicks are ready for market, he removes them to a fattening pen, the floor of which is slatted. These he feeds a meal of two parts corn and one part wheat, ground together, three pints of ground meat, and a spoonful of Douglas mixture to a pailful. Mix very stiff with hot water; let it stand two hours, and feed three or four times a day, as much as they will eat up greedily.

At first it is best to feed the chicks every two hours, all they will eat up clean. After about two or three weeks old they can be confined to three meals a day. The first feed of the day should be given at daybreak, and the last feed a little before they are ready to creep in their brooders for the night.

Great care must be taken in the preparation of the feed. It must not be sloppy, neither hot. It should be just moist enough so as to be easily crumbled, and warm.

Feed must be given in troughs. If thrown upon the floor it will be trodden under foot and wasted.

It must not be forgotten that the methods of feeding herein given are for broilers alone. In raising birds for breeding purposes, more attention must be paid to growth of bone and muscle than fat.

With regularity in feeding, and a regular warmth in the brooders, two-thirds of the troubles in raising chicks can be avoided. As brooder-raised birds are free from lice, and are never troubled with the gapes, it shows that if there is a failure in the method, it must be through the instrumentality of the man. As we have said before, good brooders and good feed are everything. It is much easier to hatch the eggs than raise the chicks.

Diarrhœa is one of the most common troubles with chicks raised by artificial means. It is caused in most cases by the chicks becoming chilled. A little of Sheridan's Condition Powder put in the soft feed is very beneficial. When the droppings stick to the feathers behind, and close up the vent, it is necessary to soften the manure with hot water, and then carefully remove it, at the same time trimming the feathers. If diarrhœa is caused by the feed,—either that it is given sloppy, or that some article in the mess loosens the bowels,—the cause must be removed before the remedy is applied.

Some chicks,—such as the Leghorns, Houdans, etc.,—on account of their rapid growth, are apt to grow their wing feathers too fast. It will make the birds droopy, as the little strength they have is robbed by this rapid growth. The best remedy is to clip off the tips of the long feathers.

When cases of colds are noticed in the flock the trouble must be corrected at once, as neglected colds have caused more disaster in brooding houses than anything else. A slight sneeze is the surest signal. There are two remedies we have found to be excellent. The one is to put a quarter of a teaspoonful of kerosene oil to a quart of drinking water. Repeat every day until cured. The other, take a piece asafœtida, about the size of a hazelnut, tie it in a small muslin bag, making it a regular "teat." Drop this in the drinking vessel, and let it remain for several weeks. This remedy we prefer to the former one, as it causes less trouble, and is more effectual. A dozen homeopathic aconite pellets in a quart of drinking water is also an excellent remedy. A few years ago the Hammonton men were taught the lesson that to neglect colds is but to invite roup. During that season several thousand dollars were lost by the ravages of the disease. Mr. James Seely, alone, lost something like two thousand chicks.

Leg weakness is caused both by too high feeding, and by too strong bottom heat in the brooder. A good remedy is to sprinkle the floor of the brooder with water, and feed more bone meal.

Overcrowding is a mistake. Pens that were built to contain fifty chicks are often made to hold one hundred. As a result, the stronger chicks crowd out the weaker ones, disease follows, and a great loss by deaths. Small flocks always pay best.

Eggs from perfectly hardy birds alone should be used. This can only be safely done when one knows the condition of the flocks from which the eggs are produced. A few years ago we made the experiment of mating a male bird that had the year previous been affected by roup, but apparently cured. We never believed that contagion could be completely cured, so we, after our bird for two or three months looked the very picture of health, yarded it with a number of healthy females. Every chick that came from the eggs laid by that flock showed a constitutional weakness, and it was with difficulty that we raised one-half of them. It solved the question so often asked: "What ails my chicks?" It proved to us that the best results alone are obtained from eggs laid by hens and sired by males that never knew contagion.

Pullets' eggs are not to be relied upon — at least, not until the pullets are ten or eleven months of age. The offspring are weak, and it requires the very best of care to raise them. But matings of two-year old hens, by a vigorous one-year old cockerel, produce the most fertile and strongest eggs. Matings of one-year old hens to two-year old cock, are also good.

In buying up eggs from any and every source, we invite disease, discouragement and failure to our brooding houses. We must conduct our business upon the strictest business principles. We must give our incubators the best of eggs, our chicks the best of feed and care, and the results will be the best of market prices.

There is no economy in buying "cheap food." The best of grain is the cheapest; damaged stuff is dear at any price. Where it is possible, it is best to buy all grain whole, and have it ground. Too often damaged grain is used when ground for poultry food.

In giving drinking water during cold weather, the chill should be taken off by adding enough boiling water until you can notice the change by the feel of your hand. It is a good plan, from the very start, to keep rusty nails or iron in the drinking vessel. It gives an iron tonic which will make the use of such preparations as Douglas mixture unnecessary.

CHAPTER V.

MORE ABOUT INCUBATORS.—THE QUESTION OF MOISTURE.—ANSWERS TO QUERIES.

As I stated in the initial article of this series, for success in broiler raising it is necessary to have an egg farm attached. This buying eggs here and there and everywhere, is a great risk. There is too much likelihood of "paying dear for the whistle." Think of thirty good eggs out of three hundred! That was the experience of one man in Hammonton who had to buy his. To lose thirty or forty per cent is not an unusual thing with broiler men who have to depend on boughten eggs.

Another reason why every broiler farmer should raise his own eggs is that he has a better chance of having stock more uniform in size, growth, etc.

For instance, here is a pen of chicks all the same age, hatched from eggs bought up all over the country. They show American, Asiatic and Mediterranean blood, not much of either, but still sufficient to tell of the mixture. Some of the chicks are shooting ahead in feathers, some are slow in growth of feathers, some show good bodies, and others are disgusting to look upon. What is the result? At killing time the farmer must pick a few out of this pen, a few out of that, and so on, instead of taking all out of one lot.

Now if he kept his own hennery and either had one variety of thoroughbreds or some good cross, he would have a more uniform lot of chicks, and less trouble to make a selection when preparing for market.

Of the list of thoroughbreds the Wyandottes lead for broiler purposes. They grow quickly as chicks, are active, take on flesh readily, and present a plump and attractive appearance. In all probability, the Langshans and Houdans come next. A great many Langshans are marketed annually from Hammonton. The old prejudice in the New York markets against white skinned birds is practically dead, judging from the amount of black fowls and chickens that are bought up from this section and at full rates. It is good it is so, for no finer meat can be had than that of the Langshan, Spanish, and some of the other black breeds. The Light Brahma is a good broiler at ten to twelve weeks of age, but the feathers on the legs must be shaved off when the chicks are dressed for market.

The majority of broiler raisers prefer crosses. They like them best for several reasons: First, they mature more quickly; second, they present more plump carcasses; third, they are more rugged — but, as to the most proper cross, there seems to be much disagreement. Among the many crosses used, the most prominent seem to be Leghorn-Plymouth Rock; White Leghorn-Light Brahma; Leghorn-Wyandotte; Leghorn-Langshan; Black Minorca-Langshan; Plymouth Rock-Brahma; and White Wyandotte-Brahma. After trying a number of matings, I have become most strongly attached to the Houdan-Brahma, or Houdan-Cochin cross; Houdan-Plymouth Rock, or Houdan-Wyandotte, are also good. I like the Houdan male better than the Leghorn, as the former is more of a table fowl than the latter, and adds more real value to the cross.

The following standard shows how to approach the ideal broiler, one that will sell best in the average markets of this country:

Weight.— From one and a quarter to two pounds each, according to the time of the year.

Head.— Short.

Comb.— Small — rose or pea preferred.

Back.— Short, broad and flat at the shoulders.

Breast.— Broad, deep and full.

Body.— Short, deep and well rounded.

Legs.— Short and stout thighs; short and stout shanks, free from feathers, and bright yellow in color.

Color of Skin.— A rich yellow, and free from pin feathers.

If broilers are sent to market in that condition they are bound to command a good sale. Large combs and feathered legs are as great objections as a poorly developed body, as they give too much an appearance of age. To make the carcasses still more attractive they should be killed by stabbing in the roof of the mouth, and then the feathers plucked while the animal heat is still in the body. Scalding injures the color of the skin, and makes a more dull look to the body. Then after being dressed they are "plumped" by dipping ten seconds in water nearly or quite boiling hot, and then immediately thrown in cold water. Hang them then in a cool place until the animal heat is entirely out of the bodies. Note how closely the above standard fits the White Wyandotte.

Some years ago a number of experiments were made in growing broilers, the object being to ascertain how much time was actually necessary to grow a chick up to the desirable weight. I believe some were brought up to two pounds in eight weeks; but it took the majority ten to twelve weeks. Even at twelve weeks the best of brooding, the best of feed, and the best of attention must be given.

Miss H. M. Williams, of Hammonton, says at ten weeks of age the chickens should weigh at least one and one-half pounds. They are weighed as removed from the pen, and an allowance of half a pound made for shrinkage. Place in slat bottom coop to prevent them from eating their own droppings, and give no water nor feed for from twenty-four to thirty-six hours before killing.

In Hammonton there are very few who do their own killing. Experts are employed — men who have become adepts in the business. They kill, dress and prepare chickens for shipment, at five cents a head. Out of this amount they pay for pin-feathering, the latter work being done principally by Italian women.

The process, briefly stated, is as follows:

The bird is fastened by the feet to a rope suspended from the ceiling. A barrel is placed directly underneath the bird to catch the blood and feathers. Then taking the bird under the left arm, its mouth is opened, and a sharp cut is made lengthwise in the mouth to bleed them, and a slot upwards to penetrate the brain. Immediately picking is begun, and it is wonderful how fast the feathers fly. As soon as the carcass is bare of feathers it is handed over to the pin-featherers, who carefully perform their work, and the bird is thrown into cold water, slightly salted, where it is kept for several hours, and afterwards hung up to drain off. When fairly dry they are ready for shipment.

Some additional facts in the preparation of broilers for market will be given in the next chapter, gathered in an interview of one of the Hammonton experts.

During the publication of this series of articles, I was fairly deluged with queries from FARM-POULTRY readers; and, while I answered all by mail, it may be a good time now to refer to them in a general way here.

One gentleman from Pennsylvania wants to know the best method of turning the eggs; and another is curious to know if turning the eggs is really necessary.

To the former we will say that nearly all the leading machines provide an extra tray. This tray is placed over a tray of eggs, and firmly grasped in the center, the trays are swung around. That is about the surest method, as every egg is bound to be turned.

Some machines have roller trays. The tray is run along a smooth board, and the rollers turn the eggs. Other machines have the eggs upon a cloth, and by means of a crank, the cloth is worked, and the eggs in consequence turned.

"Is turning the eggs absolutely necessary?" asks the other correspondent. Yes. Repeated experiments have proved that no hatch can be assured by not turning them. The hen turns her eggs — and how can we have success in artificial methods by not following her plan?

One correspondent wanted to know what is the matter when the hatch is prolonged over the time — twenty-one days, and what to do in that case. The trouble is occasioned by irregular running. A fall of temperature to ninety degrees, and kept there for several hours, will prolong it. The best remedy is, as soon as the eggs begin to pip, to wet sponges with boiling water and place at different parts of the incubator. This steam and moisture softens the shells and helps the young out.

"How much ventilation is needed in the incubator, during the hatch?" asks a lady. No rule can be given that will hold good in all machines. Some require more and some less. The manufacturers' directions are the best to go by. Ventilation, like moisture, can be overdone — and both are of the utmost importance when the eggs begin to pip. In nearly all machines, the ventilators should be open but very little in the start, and the opening increased as the hatch progresses, and wide open when the chicks begin to come out. We have now, at this writing, a good hatch coming off in the Paragon machine (built expressly for Judge J. H. Drevenstedt from plans he received from France), in which we kept the ventilators closed until the eggs were pipped, when they were gradually opened. This rule might not hold good in other machines. The only sure way of knowing the proper amount of moisture is by testing the eggs. When the air cells are smaller than those given under the head of moisture (see Chapter II.), ventilation is needed; when larger, ventilation must be stopped.

"I see spraying the eggs after the fourteenth day twice a day with water warmed to 110 degrees is recommended. Do any of the Hammonton men practice that method now?" Not to the best of my knowledge. Moisture can better be supplied by wetting sponges with boiling hot water; but even this is not necessary until the eggs begin to pip. Strictly speaking, it is not necessary to have any moisture the first two weeks. It is the last week of the hatch that makes a demand for both moisture and ventilation.

"How often should eggs be turned?" We turn twice a day.

"What objection is there to setting eggs from pullets?" None whatever if the pullets are not less than ten months old. Eggs from too young pullets may hatch well enough, but the stamina is not there to insure thrifty birds. Eggs from two year old hens are best.

"Can duck eggs be run in the same incubator with hens' eggs?" No. A duck egg requires more ventilation than hens' eggs, on account of being larger. The air cell will have to be larger, so as to afford room for the duck to turn its head and bill. The amount of ventilation necessary for duck eggs would be too much for hens' eggs.

"Will a machine that makes good hatches with duck eggs, perform as good work when charged with eggs from hens?" Certainly, if the moisture and ventilation are properly attended to. Mr. Rankin, who hatches thousands of ducks annually in his Monarch machines, fills them at certain seasons with chicken eggs, and with equally good success. I have repeatedly tested this same thing with nearly every make we have tried on our farm, and never, in a single instance, saw any difference where all other things were equal.

"I have only a few fowls, and thought of getting an incubator. As I get only a few eggs each day, what machine would you advise me to get that will run successfully by adding eggs as I get them each day?" There is not a machine in the world that will allow that. After the machine is started, there must not be any new eggs added. The new ones would chill those already started.

"How long can eggs be kept before setting — how should they be treated?" The fresher the egg, the better the guaranty of a good hatch. By placing them in an egg crate, either end up, and turning every twenty-four or forty-eight hours, we have had them keep for a month, and hatch a good percentage. Eggs for breeding must be kept in a cool (not cold) place, and perfectly dry. Eggs that have become chilled are unfit for hatching.

"Are eggs, tested out on the fourth day and perfectly clear, fit for culinary purposes?" Yes — even on the seventh day they can be used. The heat of the incubator merely ages such eggs. That is, an egg subjected to a temperature of 103 degrees for four or seven days puts it in a condition that the egg would naturally attain at twice its age. Cooks use them, and find no difference whatever, but bakers say they do not do so well in cake making. They certainly are not *fresh eggs*, even at four days old; but are to be classed with the eggs the farmer often gathers when he comes across a hidden nest (and which, unfortunately, he dumps in the "fresh-egg" crate).

"When a chick is unable to get out of the shell, is it not best to break off small particles of the shell, and thus give it a chance?" By all means, no. You can better help matters by putting in the aforesaid sponges of hot water — but never try to free a chick from the shell. The chances are your picking at them will start the blood to come, which will only end in the death of the chick. A chick unable to free itself, will be too weak to live if helped out.

"Is there an incubator made that can be run in an outbuilding where no artificial heat is kept?" All machines, more or less, suffer from changes in outside temperature. For that reason a dry cellar is the best place. An incubator run in a kitchen would vary as much as one in an outbuilding, — for during the day, when a strong fire would be necessary in the range, the temperature of the room would increase the heat in the incubator; while at night, when the kitchen fire could be dampered, the temperature in the incubator would fall; but in a dry cellar, there would be very little change, and scarcely enough to vary any in the incubator's heat. The broiler men who have incubator houses attached to their brooding house, are compelled to watch the machines day and night. In fact, during hatching season, the broiler man is continually at his post, to meet any variation; not a few of them have hammocks or couches in the incubator room, and take hour naps the entire night.

In artificial methods, it must not be forgotten that, in regard to the application of heat, success in incubating is measured by the regularity of the temperature. That is, the closer the desired temperature of the incubator can be kept the entire period, the better will be the results. And in brooding, keeping the chicks comfortable at all times, is the safest rule to go by. Too much or not enough heat for the chicks will meet with as much loss as variations in the incubator will occasion.

Regularity in heat and regularity in feeding, are two important matters in artificial methods, and those who work on that line meet with success. Get a system to your work, and follow it out carefully.

CHAPTER VI.

DRESSING AND SHIPPING TO MARKET.—SEVERAL VALUABLE QUERIES.

As stated in Chapter V., the killing and dressing of broilers for market is done principally by experts. The birds are dry picked—by that is meant they are never scalded, but plucked while yet the life warmth is in them. It is wonderful with what dexterity this work is performed by those who make it a business.

In order to give a clearer idea of the work than what has been heretofore published, we interviewed one of the Hammonton expert dressers, Frank Y. Hopping—a young man who some years ago managed the large broiler plant of Mrs. Judson, of Ardsley, New York, (daughter of Cyrus W. Field). Mr. Hopping could not give an exact average of birds he could pick and prepare for market in a day, so much, he said, depended upon the condition of things at the plant at which he was doing the work; but the number would run somewhere between one hundred and fifty and two hundred birds.

The birds prior to the picking are caught and weighed, and those that are up to the desired weight (one and a half, or two pounds, as the case may be), are put in a cage, and those short of the weight, are returned to the pens and held back for another week. He takes the work on contract—five cents each, out of which he pays the pickers (Italian women) two cents apiece. These women average from fifteen to fifty birds in a day, their work being to take out the pin feathers, so as to have the carcass perfectly clean. When it is known that these women must pick out every little stub, done with the fingers and a small knife, and all for two cents a bird, their work can certainly be appreciated.

Everything being ready, the bird's legs are fastened to a stout cord suspended from the ceiling, and a large hogshead or barrel is placed underneath to catch their blood and feathers. Then the operator gets in front of the bird, placing it under his left arm; and with a knife made expressly for the purpose (sold by dealers in poultry supplies), he runs the knife back in the mouth, and then bringing it a little forward cuts cross-wise, severing an artery. The mouth, during the operation, is held open with the fingers of the left hand. Great care is taken not to cut too much, for fear of the bird dying before the feathers are all removed, in which case it would be difficult to pick.

While the life blood is still in the chicken the rapid work of feather pulling is begun. The feathers of the breast are first taken, then the neck, then the back, then the tail and wing feathers, and finally the feathers on the legs. It is a sight to those who never saw the work before. As if by magic every touch of the operator makes the feathers fly, and before one can realize it the bird is perfectly clean, excepting what pin feathers may remain. As soon as the long feathers are pulled the women begin their work, and before the carcass has a chance to get cold, it is as bare and clean as it is possible to make it.

After the women have completed their part they hand the bird back to the dresser, who gives it a critical examination before it gets the first bath of cold water. If the skin should be torn, which occasionally happens, it is sewed up with common thread.

Having all the feathers removed, the birds are then put in cold water, to which is added a little salt. After remaining in this water for some time, the clotted blood in the mouth of the chicken is removed with the finger, and the carcass is placed in another tub of clean cold water.

This ends the work of the dresser, the shipping being done, generally, by the owner. During cold weather the birds are packed in barrels or boxes, one on top of another, until full. When warm weather arrives a layer of cracked ice is put in the bottom of the barrel or box, another layer of ice in the center, and a layer on top.

"How many incubators and brooders will I need, to run the broiler business? I am willing to commence the business on a small scale, and grow up with it." Supposing that you are a novice at the business, I should advise to begin with one incubator, and two or three separate brooders, and thus gradually establish yourself in experience as well as facilities. It never pays to undertake more than you know how to handle, and as experience is as necessary as cash in the broiler business, it is best to begin small, and "grow up with it."

"We have been trying to operate two incubators made after the *Poultry-Keeper* plans. Both machines are in a cellar room under the house kitchen. The cellar is 13 x 14, with a sand bottom; in one side of this room, about three feet from the machines, is a well covered with a wood platform, and connected with the kitchen pump by an iron pipe. The cellar is what would be called a dry cellar, were it not for the well of water in it. The water is about six or seven feet from the floor level. The temperature is 55 or 60 degrees. Now let me give our hatches: The incubators hold two hundred and forty eggs each. First hatch, sixty-five; second, forty; third, thirty-three; fourth, twenty-six. Good record, isn't it? Now we follow the directions exactly, and found but very little trouble in keeping the heat at 103 degrees. We used no moisture the first week. The second week we put in three sponges the size of an egg, and wet them twice a day when we turned the eggs. The last week we put in six sponges. We found lots of dead chicks in the shells, just ready to come out. Was too much moisture the cause? How much would you advise? Is 55 degrees too low a temperature to turn the eggs in? Do you think the well has anything to do with our poor success? Would cementing the cellar bottom make it any better? Would it be advisable to have a stove in a cellar of this temperature?"

Undoubtedly, too much moisture is the cause of the whole trouble. The manner of applying the sponges — none the first week, three the second, and six the third, would be about right if the machines were run in a room above ground; but that well of water, and the more or less moisture which would naturally be in a cellar, are supplying considerable moisture to the machine. Our plan would be, no moisture the first two weeks, and six sponges the last week. And when the eggs begin to pip, I should have the sponges soaked in boiling hot water. Too much moisture grows the chick too fast in the shell, or rather swells it up in the shell so much that it is unable to turn and work its way out; consequently it must die. This seems to be the trouble with the correspondent's hatch. Fifty-five degrees is not too low a temperature to turn the eggs in, provided the work is done quickly; but I should not give the eggs an airing any more than just what they get in turning. The fact that our correspondent had no trouble to keep the temperature up to 103 degrees, shows that there is no necessity for any artificial heat in the cellar, as the temperature must keep very regular.

"I have read with interest your article in FARM-POULTRY for May. It was especially interesting to me on two points in particular: moisture (that perplexing question), and turning eggs. You ask, ' How can we have success in artificial methods by not following the hens' plan?' Well, then, why should we have to make such a departure in the moisture business? Is there any more moisture under a hen the nineteenth day than there was the first day? Now as to the absolute necessity of turning the eggs, that same question has been pretty much in my mind for some time, but have had no opportunity of thoroughly testing it. You say, ' the hen turns them.' Have you any idea what she does it for? If we are obliged to turn them, I think it would be so much better if we could do it intelligently, than by a mere mechanical movement of the hand."

If we knew for a certainty the amount of moisture that was under a hen from the first to the last day of the hatch, there would be no further need of experimenting on that score. When we build a machine to take the place of the hen in hatching, we can pretty nearly imitate the hen in every way but moisture. In this, then, we must experiment. The quantity the hen supplies we do not know. We could not even say she supplied any; but it is necessary to give more or less when we run a machine. More if the *outside* temperature is less; less if the outside temperature is more. In other words, with an incubator a certain amount *must* be given. What is lacking in the room in which the incubator is placed, must be given in the machine, and *vice versa*. This fact has been proved by running a machine in a dry atmosphere in a room above ground, and the same machine in a cellar. A hot water machine can be run in a damp cellar with tolerably good success, without any moisture being supplied the entire hatch. This has also been proved.

Why the hen turns her eggs, is about as hard to answer as to solve the cause of her cackling after laying an egg; but that she does turn them, there is no doubt. And from the fact that she turns the eggs, it follows suit that nature demands it, and we copy after her. The experiment has been repeatedly tried in Hammonton, but never with any success where the eggs were not turned.

"What is the correct temperature of heat for hatching? Some say 102 degrees, and some 103." Any heat from 95 to 105 degrees will hatch. Of course, a hatch continued long in the nineties will not be practically due in twenty-one days, but will run several days over the time. The more regular the temperature can be kept at 103 degrees the better the chances for success.

"Is there any truth in the assertion that fowls that were hatched and reared by artificial methods are not as good for breeding purposes as those raised by natural methods?" I have never been able to notice any difference. There can be no truth in such an assertion. How incubation, either natural or artificial, can affect the stamina of an egg, I cannot understand.

"My chicks (Leghorns) grow their wing feathers so fast that they almost drag, and it makes them droopy, some even dying. What should I do?" It is a good idea to clip off the ends of the wing feathers when they grow so fast as to interfere with the thrift of the bird.

Here are a few notes, suggested in this month's pile of correspondence:

Never put cold water in the moisture pans, as there is danger of chilling the eggs.

Never remove the chicks until perfectly dry; it is better for the chick and the unhatched eggs.

Never test the eggs in a temperature less than 60 degrees. Testing can be done any time after the fourth day. I prefer the seventh and fourteenth.

New milk must not be given to chicks. It should be diluted with nearly its quantity of boiling water.

CHAPTER VII.

MORE QUESTIONS AND ANSWERS. — HENRY NICOLAI'S BROILER PLANT. — A GERMAN'S PROFITABLE COMBINATION.

A subscriber from Gustavus, Ohio, writes:

"We wrote you in regard to our unsuccessful hatches with our incubators. On receiving your reply we at once re-set our machine in the same cellar we described to you. The weather has been so very rainy that we thought even less moisture than you said might be about right, so we used none until the eighteenth day, when six sponges were put in. On the nineteenth day a few eggs were pipped, and the sponges were wet in boiling water. The drawer was kept closed after that until the twenty-second day, when we took out thirty chicks. The thermometer was down to 99 degrees for about two hours twice, and never above 104 degrees, and almost always at 102 and 103 degrees. On examination we found two-thirds of the eggs contained chicks full grown, but dead. In all we have made six hatches from fifteen hundred good eggs (eggs that are ninety per cent fertile, and that hatch well under hens), and from which managed to blunder out two hundred chicks. We have tried every plan possible, and have one hope left, and that is that our thermometers are wrong. We mail them to you this day, and will be very grateful if you will test them."

The thermometers arrived, and the first glance told the cause of much of the trouble. The mercury was parted in both of them, and placed under the wing of a broody hen there was no agreement of temperature. There was at least ten degrees difference in the two.

It is remarkable how well they did in the incubators. A great many of the failures to successfully hatch can be attributed to poor thermometers; and it is very important that reliable ones are used. Regularity of temperature is the great desideratum, and the more closely it is kept the better will be the percentage of hatch.

"We have two broods of chicks, one hatched March 20, the other April 20. From which would you save pullets to commence laying by November? They are Light Brahma and Indian Game cross. If the younger would do just as well I would rather keep them, as the older ones will bring a good price just now."

Your April hatched birds should begin laying by the last of November. The cross should make a good combination for broilers.

"What is the best plan by which to build a brooding house?"

The houses most generally used have large sashes of window lights on the slant of the roof. The adopted plan of the future, however, will do away with this part glass roof, and instead there will be half-window sash in *front* of the house the same as is used in henneries. In fact, a good brooding house is one fashioned after a substantial hen house. It must be remembered that the house does not perform the most important part of the business—a good brooding system is more imperative — a mere shed answering if there is sufficient provision made to shield the young from the weather. The healthiest looking flock of chicks I ever saw, was owned by the Misses Pressey, of Hammonton. Their houses are really nothing more than open sheds protected by cloth instead of glass. In our interview with Mr. Pressey, father of the young ladies referred to, we will give a better explanation of their methods.

The broiler farm of Henry Nicolai, of Hammonton, is one worthy of study. It is not a large concern, as regards building, yet it is large enough to yield its owner an average of four hundred dollars a year profit. The brooders are heated with the pipe system bottom heat. The incubators used are run exclusively by hot water, and were built by Mr. Nicolai himself. He prefers hot water, as in his experience it brings out much better chicks, and is safer. The first thing he gives chicks, after forty-eight hours old, is cracked wheat, and gradually he gets them on to whole grain. He says he has long since given up the various mashes and cakes advised, and his stock look none the worse for it. A noticeable feature of Mr. Nicolai's house is cleanliness. He finds that it pays to keep down the cobwebs, and have the house generally clean. He is a firm believer in fresh air; and allows his chicks out in their small yards on all nice days. Mr. Nicolai began at the bottom of the ladder. His first machine was a small one, and his brooders were crude ones. With these limited facilities he began his profitable plant.

In the town of Hammonton lives a German who I believe is following out the right course. He has a brooding house and several incubators. I understand he annually raises about one thousand broilers. He also keeps cows and sells milk; and grows quite a lot of vegetables for market. He may not be coining money, but I venture he has as independent a livelihood as a man could wish to have. Would he do better if he should make broiler raising an exclusive business? I doubt it. That there is a big profit in the business, no one can deny; but to give up everything else and devote your entire time, there is a risk. I say these things not to discourage those who contemplate embarking, but rather to steer them on the right course. While I believe poultry in any branch does better as an adjunct, I do think that an exclusive poultry farm would pay where every branch is handled—for instance, where a man raises broilers, roasters and eggs, by selling the eggs when the prices are high, and hatching them when there is a decline in the market.

In France the raising of chickens is a most important industry, says Mr. Camille Michel, a resident of Hammonton—who came here from Nantes, France, several years ago. He said that the cottagers there make it a part of their livelihood to furnish choice plump broilers for the tables of the rich. The Houdan, Crevecour, La Bresse and LaFleche breeds are more generally used; but in the town of Houdan the Houdan fowl alone gives the choice birds, and that fowl has more friends in his country than any of the French breeds. Mr. Michel further said that in the production of broiling and roasting fowls they use hot water almost exclusively, and that the average Frenchman could not be persuaded to try any other method. When questioned about our plans of operation, he said he thought the Americans are ahead of them in some things, but in others France could teach them a lesson. For instance, the cottagers there do the work as an adjunct to other work. They do not try to make it an exclusive business, except in some cases. He thought that if our people would raise less carcasses and choicer meat, there might be a healthier growth to the business. Here quantity, and not quality, seems to prevail. The broiler men are evidently living in the belief that an egg is an egg. It seems that they do not care what that egg will hatch, only that a broiler is made. Mr. Michel has hit the nail upon the head. If each broiler farm would raise its own eggs, and the stock from which those eggs are secured would be either a desirable thoroughbred or a systematic cross, there would be a more uniform growth and more profit. Look at a pen of growing chicks hatched from eggs bought up here and there, and you will find all sizes and conditions. Instead of the uniform lot that a good cross would make, there is as much difference in them as would be perceptible in chicks of different ages. It is a common thing for the dresser to sort out the weights from pens differing in ages. This would be unnecessary if all were of the same blood; and the individual shipments would be larger in consequence.

Another important item in the growing of eggs is the knowing of the condition of your stock. Eggs from inbred or sickly parents certainly cannot produce hardy and thrifty offspring. Even Mr. Jacobs, who formerly believed that it paid the broiler men better to buy their eggs than to bother with hens, now says: "As long as operators buy their eggs for incubation, the work will be uncertain."

A reader of FARM-POULTRY wants to know if the cause of chicks dying in the shell can only be attributed to too much moisture? No, there are other reasons. Too high or too low a temperature, or a lack of constitutional vigor or condition, are common causes.

"My chicks are doing very nicely," says an inquirer, "but I notice those that are growing their feathers very rapidly are troubled with diarrhœa. What is the cause of it, and what would you advise me to do?"

The trouble is due to a lack of nitrogenous matter in the production of feathers; and raw meat should be fed once a day. Chicks that have a range where they can pick up worms and other insects are never troubled that way. Such breeds as Leghorns, Minorcas, Games, and Spanish, are in need of more meat than any of the American or Asiatic classes, as they grow their feathers more rapidly.

"I am told that you people in Hammonton are never troubled with the gapes in your flocks. What do you do to prevent the trouble?"

Nothing. Our soil is white sand, which is very porous. All accumulated filth is carried away after a hard rain.

When the broiler season in Hammonton closes, the yards of the brooding houses are converted into miniature gardens. Being richly fertilized by hundreds of chicks, the broiler men raise almost enough vegetables for their own table use in the space that would otherwise lie idle until next season. After gathering their crop, some of them sow rye, and thus purify the runs. The finest tomatoes and corn that I ever saw were grown in such yards, without any additional manuring. Plum trees do well in the yards, and besides an increase of fruit, the shade is beneficial to the young stock.

CHAPTER VIII.

AN INTERVIEW WITH P. H. JACOBS.— QUESTIONS ANSWERED.

Those who know P. H. Jacobs, of the *Poultry Keeper*, know that it is a hard matter to interview him on the broiler business for publication in another paper; but I took him unawares. Mr. Jacobs told your correspondent that the first broilers were raised in Hammonton in 1882. It was an experiment at first, but by careful study it developed into a paying industry, and soon the excellence of the Hammonton product became famous in the Philadelphia and New York markets. In a comparatively short time the fever spread, and brooders were erected all over the town. An association was organized, and everything flourished until jealousies arose, when the club disbanded, and each man ran matters to suit himself.

This local mentioning is made in reply to a query from a FARM-POULTRY reader, asking: "About how long ago is it since the first broilers were raised in Hammonton?"

I asked Mr. Jacobs what he estimated the cost of raising broilers up to one and a half or two pounds. He replied that his experience proved that five cents per pound was all that it costs, although in some cases six cents was reached, but never over that. They double their weight every ten days until forty days old. I asked Mr. Jacobs what he considered the best cross for broilers, and he said Plymouth Rock crossed on Brahmas. This cross, he said, is probably not the best from a strict meat judgment, but it gave the best satisfaction to the broiler man.

"What do you consider the best feed for young chicks?" I asked. He thought the question was a hard one to answer, but for the first week he knew of nothing better than rolled oats. Pin-head oatmeal was a favorite with Mr. Howe; but in Mr. Howe's broiler days rolled oats were unknown. He believed the latter far superior. After the first week wheat and cracked corn should be given. Of course there are "side-dishes" which must not be overlooked, such as bone meal, grit, charcoal, green stuff, etc.

Then came the stunning question, "Which is your preference — hot water or hot air for artificial methods?" No wonder Mr. Jacobs hesitated. An editor is very often condemned for saying what he believes to be true, and his opinion was narrowed down to: "For brooding, hot water is the best for wholesale work; and hot air for small houses."

"How about incubators?" I asked.

"Well, I like hot water the best."

"Now, Mr. Jacobs," I continued, "do you favor broiler raising as an adjunct, or as an exclusive business?"

"Both. When operations are only to be continued in the winter, I say adjunct; but where the house is to be run the entire year, it can be made an exclusive affair."

"Do you think it pays to hatch during the summer?"

"Yes. During last year, for the month of January, the lowest price received was twelve cents, and the highest was thirty-six cents. In February, twelve the lowest, and thirty the highest; March, twenty for lowest, and forty for highest; April, twenty for lowest, and fifty-five for highest; May, twenty-eight for lowest, and fifty-five for highest; June, twenty-six for lowest, and thirty-seven for highest; July, twenty for lowest, and thirty-three for highest; August, eighteen for lowest, and twenty-eight for highest; September, sixteen for lowest, and twenty-two for highest; October, thirteen for lowest, and twenty-two for highest; November, thirteen for lowest, and twenty-two for highest; December, thirteen for lowest, and twenty-one for highest. These, of course, are pound prices. It will be seen that the lowest quotation is fifteen cents, and that price occurs only in the months of October, November and December. Another point, and one which will strengthen my case, is that during the months from about the middle of February to the middle of September, the prices call for broilers, usually one and a half pounds each, while the rest of the year they are *spring chickens*, and may weigh two to three pounds, bringing a good price at the lowest quotation. The cost of artificial heat during warm weather also lessens the cost of production."

This interview has brought out a number of suggestions, and it may be well to make a few notes and comments.

Mr. Jacobs places the cost of feeding up to a marketable size, at five or six cents a pound. Mr. Seely thinks the estimate I gave in my first article, fifteen cents a pound, which includes brooding, about right; but the matter of cost can only be reckoned by taking the wholesale figures on grain at the nearest market. However, in making up an estimate it is best to place the price of feeding at fifteen cents per pound, including the cost of brooding, or ten cents a pound for feed alone. The cross Mr. Jacobs recommends is one which Mr. Phillips endorses as the best. Mr. Phillips, as we have previously stated, conducts the largest house in Hammonton. A good point is made in favor of summer hatching, by the quotation of New York market rates. Where chicks have an unlimited range on a farm, the cost would naturally be lessened, and as "spring chickens" can weigh as much as three pounds each, there would be a profit even in the low rate of thirteen cents. This is a matter that should have some experimenting.

How different the market of today is from that of about ten years ago. Then January, February, March and April were the broiler months; while today, according to Mr. Jacobs' statement, they are "broilers" from the middle of February to the middle of September, showing an increase in demand. It is reasonable to suppose that ten years from now there will be a demand extending the entire year. Ten years ago the broiler size was one pound; now three pounds to the pair is the way they are generally sold. The prices then were sixty cents to one dollar a pound; now they bring from thirty to sixty cents per pound.

In 1884 Mr. Jacobs made a calculation, which I think hits the mark of cost better than what he claims now. He then said: "We have shown that it costs but ten cents to keep a chicken ten weeks." One cent a week for a chicken, and two cents a week for a fowl, are generally accepted estimates.

In the same year Mr. Jacobs makes an estimate, or rather gives a table of profit and loss, what an incubator will pay in three months, as follows:

EXPENSE.

Cost of incubator	$25.00
Cost of one hundred eggs, three cents each, four trials	12.00
Oil	3.00
Feed for one hundred chicks	10.00
Four brooders, at $10 each	40.00
Total expense	$90.00

RECEIPTS.

One hundred chicks, at seventy-five cents each	$75.00
Cost of feed, oil and eggs	25.00
Profit	$50.00

"The incubators and brooders being permanent investments, their cost should be proportioned among all succeeding hatches."

I give this estimate by Mr. Jacobs because it shows in a measure how profitable an incubator and a few brooders on a farm would be.

Every now and then applications come in my mail for "jobs on a broiler farm," and equally as many letters are sent me asking for good experienced men to manage such farms, — and I cannot help either. This may seem a little strange to the average reader, who would naturally suppose it to be an easy matter to put the applicants on the road to the open jobs. But when you come to consider that nine-tenths of those who want work on broiler farms are men of little or no experience, who have only an experience in natural methods, or who have become the victims of "windy" boom articles, and on the other hand those who want men to "run the farm" have no practical experience, and likewise have been led into this business "on false pretenses," I say it is a hard matter for any one who wishes to do justice to all concerned, to recommend or advise. Therefore, it is useless to apply. The best advice I could give would be to apprentice yourself to some practical man, and then let the employer give the recommendation.

A party in Decorah, Iowa, writes:

"We are about to start a broiler plant, and would like to ask you a few questions: First, we will tell you of our surroundings. Our marketing place is Chicago; in broiler season we can get from seven to ten dollars per dozen. Then we can buy feed much cheaper than in the eastern states. We are thinking of renting a large brick building that has been used as a brewery, which is very light and warm, at a small rent. We have given our attention to chickens for four or five years, raising with hens, but we have had no experience with broilers in winter. We have three incubators, and will give them all our time until winter getting the experience. So far have had no trouble; they work fine. Now what I want to know is, do you think we could be safe to invest in brooders and such things necessary to run the business? Second. Don't you think, considering feed and market, that our profit would be greater than that of the people who are in the business in the east, and we would be safer in running a little risk here than there? Third. Don't you think the building would be a good one for that business?

"We have read your articles with interest, but they have discouraged me a little, so I thought I would write you and get your opinion and advice before starting, stating prices of feed, and chickens when sold. The building has large dry cellars, good water, and sheds that would make good scratching pens. After broiler season, up till September, we can get from fifteen to eighteen cents per pound for spring chickens. Don't you think it would pay us to raise chickens the year round, doing nothing else? We have eighty laying hens and three hundred young chicks; we buy all the feed, and it only costs us sixty cents per week to keep them all. This will give you an idea what our feed costs. We give to the young corn meal, wheat, oats, and bran mash. The cost of shipping is not very heavy from Decorah to Chicago, as it is not far. Price of feed by the lot: Oats, 80 cents per cwt.; shorts, 75 cents per cwt.; bran, 75 cents per cwt.; wheat, very best, $1.20 per cwt.; corn, very best, 55 cents per cwt. The Chicago market, July 5th, was: Eggs, 15 1-2 cents per dozen; hens, 10 1-2 per pound; springers, 17 to 18 cents per pound."

For convenience I will answer by numbers:

First. It is safe for anyone to invest in a broiler plant if he is adapted to the work, and willing to begin at the bottom of the ladder. It is certainly a foolish thing for anyone to jump into the business without being properly equipped with such facilities as energy, ambition, experience and cash. The first two requirements must be to the fullest extent, and as much of the latter two as possible. Where there is a limited cash and experience, it will be a regular suicide to undertake too large a concern. But in the case of our correspondent, with enough laying hens to supply the eggs, four or five years experience in the feeding and caring of chicks by the old method, and a preliminary education in the running of incubators, I say go ahead, but go slow at first, and do the work well. I feel sorry if at any time I have discouraged any one, for such is not my mission. I have the greatest faith in the broiler business; I have always said that, for the money invested, it paid a good dividend; and that it certainly was a good paying job to those who handled it correctly. I want to caution rather than discourage; I believe that as a business it has been unmercifully boomed, and that such booming has been unhealthy.

Second. No; I do not think there is more money in it in the west than in the east, notwithstanding the fact that feed is so much cheaper. While the prices quoted by our correspondent are very good, those received by the eastern men are better — but taking the cost of production in both sections, and subtracting it from the receipts, I think there would be an equality.

Third. I do not think much of a brewery, or in fact any building short of a regular brooding house, for the successful raising of broilers. The mentioned cellar would be just the thing to run the incubators in; but without seeing the remainder of the building, or not knowing anything of the plans to be adopted, I am not favorably impressed with the idea.

Fourth. The broiler men of Hammonton do not have any faith in summer operations. They have tried the plan, and claim that there is more loss than in winter, and that the prices obtained do not justify this loss. Probably in the west it might be more profitable.

"How do you tell the sex of chicks that are quite young? What are the certain markings?" With the majority of breeds it is hard to tell the sex of chicks until they become a month or two old. In quite young chicks it is impossible; at least, I have never known of any such secret.

"Do you caponize broilers? Is caponizing a success financially?"

No; broilers are never caponized, although Dow says they can be. That is, he says when they have arrived at the age of two months, cockerels may be performed upon, provided they are of sufficient size. However, other experts say from three to six months is a better age.

"How much space is needed in yard for twenty-five chicks almost ready to sell?" A yard 3 x 6 feet would do.

"If you wanted to keep one hundred hens of common stock, slightly mixed with Rocks, Wyandottes, Cochins, and Brahmas, what kind of males would you get for next year? I want for both broilers and eggs."

Use Brown Leghorns.

"I have an ———— incubator with which I have had a good deal of trouble in getting the temperature the same in all parts. I had to raise the two end drawers one and a quarter inches higher than the two middle ones. Do you think it would hatch all right rigged that way? Is it common for all makes of incubators to vary in heat in different parts? One hundred seems to be as many as I can get out of this machine. Is that a good hatch?"

If the machine is standing perfectly level it should evenly distribute the heat. If it does not, close up the draught holes at those parts where there is an insufficient heat. In case this does not correct the trouble, carefully examine the entire machine, and you may find some of the parts warped. The plan of raising the drawers may accomplish the object, but I doubt it. There must be an even distribution of heat to guarantee success, and that is a point the manufacturers of all machines make. The average hatch is fifty per cent.

"Seeing by your articles in FARM-POULTRY that many of the poultrymen there use the hot water incubators, I take the liberty of writing you my experience, and would esteem it as a favor if you would reply. I made one of the Poultry Keeper incubators from the plans in P. H. Jacobs' book, 'Incubators and Brooders.' I had no difficulty in keeping the heat up, and went strictly by the rules in the book, which says: 'Keep the heat inside the egg drawer at 105 degrees the first week, 104 degrees the second, and 102 degrees the third.' Once or twice the first week the heat got up to 106 degrees, but only for a short time. The second week it was kept at from 103 to 104 degrees, and the third at about 102. It did not go below 101 degrees, nor higher than 103. I used no moisture the first week, but during the second I kept a box of wet sand in the ventilator under the egg drawer, and the last four or five days put two tin cups in which was a small sponge kept wet in the egg drawer itself. I turned the eggs night and morning. Out of one hundred and five eggs I had about eighty good fertile ones, and when I tested them about the fourteenth day, most of the germs were alive, and could be seen to move. I was looking forward to a good hatch. On the night of the nineteenth day three of the eggs were pipped, and the next day they died in their efforts to get out of the shell. I got in all six live chicks, three of which I killed after keeping three days, they then being too weak to stand. The rest are all dead in the shells except what died in their efforts to get out after the egg was pipped. The skin under the shell of the eggs was dry and hard, and some of them had to be helped out. I never ran an incubator before, but from what I have read, I thought that the condition of the eggs showed a lack of moisture, but I had more than was recommended in the book. The machine is kept in my kitchen, where the heat ranges from 60 to 75 degrees, and is not more than eight feet from the stove. I thought myself that I had a little too much heat the first week, and not quite enough moisture. Thought that possibly I did not keep the sand wet enough. I am certain the fault was not in the eggs, for they are hatching tip-top under my hens. I got thirty-eight chicks out of thirty-nine eggs about four days ago."

I republish the entire letter, as it rehearses the first experience of the average amateur. The trouble might be assigned to two causes: In the first place, a kitchen is the worst place in the world to run an incubator. The changes of temperature are too sudden, and as a rule, too great. If instead the machine had been placed in a moist cellar, the amount of moisture given the hatch would have been nearer correct. As it was, there was too little given at the proper time. I know of grand hatches made by the Poultry Keeper incubators when liberally supplied with moisture; and I have known of good ones made without a drop of moisture being added the entire hatch; but of course, in the latter case incubators were kept in the cellar. It is my opinion that the high temperature in the beginning of the hatch had a good bit to do with the poor hatch. I have never had such good returns as when I kept the temperature at 103 degrees. Opinions may vary as to the correct number of degrees necessary, but it matters little as long as the heat is anywhere regular. If you have no cellar room for the machine, it is best to double up the quantity of moisture the third week, as it is plainly seen that the given amount was insufficient.

A correspondent from Reading, Pa., thinks the pure-bred man is not in for a boom among broiler writers, and wants to know why a thoroughbred fowl is not as good as a common dunghill for market purposes. A thoroughbred fowl certainly is, not only as good as a dunghill, but far better; but with the exception of probably two or three, the average pure-bred, like the common bird, does not combine both plumpness and early maturity. A good cross will give this. Just what is the best cross to make is a question among broiler men, the desired results being obtained by mating heavy bodied hens with light bodied quick growing males. That is about the only object. I do not believe in common fowls for this business, as the past experience of Hammonton men has been that common stock gives such an irregularity of sizes and weights.

CHAPTER IX.

The Model Brooder House.— Hammonton Criticised.— Experiences.— Fattening Broilers.— Questions and Answers.

In the August, 1892, issue of FARM-POULTRY, Mr. Hunter gave what he called his "Impressions of Hammonton." Unfortunately he visited the town during "strawberry season"— a time when all the brooding houses are closed, and the broiler raisers generally are busy with fruit and vegetable culture.

These "impressions" of Mr. Hunter, while generally correct, dealt the town a blow that it did not honestly deserve. Hammonton started the broiler work in this country, and paved the way for successful plants all over the United States.

It cannot be denied that Hammonton for this industry has been greatly over-estimated, and that in many ways her broiler men are "behind the times," yet their efforts have created a substantial boom in the direction of hatching and brooding by artificial methods. They are not very progressive, I admit, but they have furnished material for others to work upon. As one of the Hammonton broiler men remarked to me yesterday, "I had a man from Long Island visit my place, and he carefully examined everything about the business, said he gained a number of valuable pointers, and went home and greatly improved upon my methods." This is but a single instance of the kind. I could mention a number more. Even if our people know better, they seem to hang on to the maxim, "Let well enough alone." Take, for instance, the style of our brooding houses. I have always held that the present hot-bed fashion was a mistake, and have thus expressed myself to several, and they agreed with me. Too much glass is a drawback. What lets in the heat during the day also admits the cold at night. Mr. Browning, of this place, who probably is one of the best booked men on artificial methods of any in our town, provided against this glass roof trouble by having window shades attached, and these he pulled down at night. It certainly was an improvement; but a better improvement, in my opinion, is no glass roof at all.

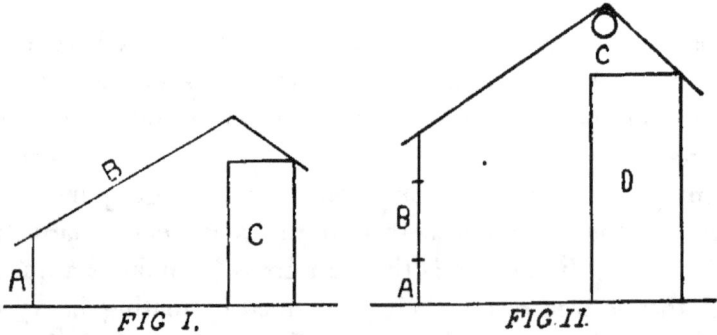

I herewith present an outline illustration of the style of house at present adopted by the Hammonton men, and a plan of what I contend is a big improvement. I feel sure that if the new plan is adopted there will be better success. A few words will explain my idea: Fig. I. is the present style. A is the front, usually eighteen inches to two feet high. At one end of each pen is a small door which lets the chicks out into their small yards. I do not like these small doors, as there is invariably a rush, jam, and some are hurt in the squeeze in their endeavor to get out. B is the glass roof, made by using the regular hot-bed sash. They are made to slide, opened or closed by the aid of a stout cord worked on a pulley. C is the entry to the building. Now in Fig. II., A is the "outlet," or, in other words, a wire netting door, extending clear across the width of each pen. The size netting used is a one-half inch mesh, and on the back of it is tacked muslin. The muslin admits enough cool air to purify the atmosphere on the inside. As these doors extend clear across the pen, when opened they at once give ample space for the chicks to get out without crowding. B is the glass window to each pen. The ordinary half-sash, as are used in dwellings, are to be used here. Either one or two of these half-sashes can be put in; but one is all that is necessary. As it has been proved that only the best chickens are raised in the fresh air, the only mission the glass can have is to furnish light; and one window to each pen is ample for that. C is a ventilator hole. The ventilator in the present Hammonton broiler house has been done away with, and well it might be, for it was an attachment to the roof of the building, which sent down the air in streams, causing much sickness. "How shall we ventilate?" has been answered to my satisfaction by having holes, about six inches in diameter, at each end of the house, up near the peak of the roof. By the use of these ventilators the air is kept more pure, and I do not notice that sickening scent so common in houses in the morning, when opening up, as is found in houses so built that it is impossible for any pure air to enter. D is an entry, same as in Fig. I.

Attached to the present Hammonton brooding house is an incubator room. Past experience has proved that such rooms are not the proper places for incubators. The air in these rooms is entirely too dry, made all the more so by the heat from the large stoves in them which furnish the warmth to the brooders. For this purpose the incubator cellar will be more generally adopted in the future. This cellar I have already described in these articles, and so need not repeat it here.

Coming back to your "Impressions," Mr. Editor, I quote from your article the remark made by the New York commission man you refer to: "You can't make a first class broiler out of scrub stock; the clean yellow legs and skin, and plump breasted body isn't there, and the result cannot be the best." It is a fact that any and everything is sent out from Hammonton. But Hammonton is not alone to be blamed for that. If the New York trade wants yellow legs and yellow skin, why don't it pay more for such than the dark legged and white skinned stock? It is a fact that fully one-half the broilers sent to New York are white skinned birds, and there is never an objection made to them, and the price is just the same as for yellow skinned stock. Certainly, the most attractive looking broiler, and probably the ideal one, is a plump body and yellow skin and legs. But the failure of the New York market to discriminate between the yellow and white skinned product, has caused Hammonton to become careless in this matter, and in consequence, her broiler men are buying up their eggs here and there, no matter what will be produced. Is it a good plan? Decidedly, no. It not only hurts the business, but it is costly to the operator. If the eggs from only one class of fowls or crosses were used, and the breeding stock fed and attended to to insure better fertility and better results, the broiler man would benefit his business at least fifty per cent. It is not an uncommon thing for the man who is picking out his stock to dress for market, to select from several pens. A broiler should be made in at least fourteen weeks, yet they are often shipped at eighteen and twenty weeks of age, for the reason that they were so slow at growing. Of course this extra age, while it does not hurt the value of the bird as far as use is concerned, has been an extra expense to the grower, as he was compelled to feed it four to six weeks longer than a quick growing bird would have required.

By mating up good breeds, such as Houdans on Cochins, or Leghorns on Wyandottes or Plymouth Rocks, for instance, and only using the one cross, the chicks would be uniform in color of skin, and would average the same weight at the same age. Besides, the cross would give quick growing and plump carcasses. It will pay any broiler farm to run its own egg supply on that principle. It will not pay on any other.

By running an egg farm in connection with the broiler plant, the owner can greatly help the business by selling eggs when the market price is high, and turning out broilers when the egg price drops. In this way the cream of prices can be obtained.

J. E. Watkis is a young man of Hammonton who combines poultry and squab culture with fruit and general farming. In a conversation with him one day on the merits and demerits of broiler farming, he replied that after he understood the principles of the business he met with good success; that he considered he was making as good if not a better dividend than from any other investment on his farm. "But," he added, "it requires eternal vigilance." So it does, and one reason why so many have gone out of the business in Hammonton and elsewhere, is on account of the close watching the occupation requires. As mindful as the old hen is of the wants of her young, so must the broiler man be of his charge. He must see that they are at all times comfortable, and never allowed to be in want. Deprive a chick of its wants and you are "pulling your own nose" by it.

Alfred Reed combines broiler raising with general farming, and finds that it pays him to arrange his hatches so as to make a shipment each week. Each week a hatch is started, and each week a hatch is due. After the first hatch has reached broiler size the second one is only a week behind, and so on. The idea is an excellent one, as it makes a weekly income to the plant. He raises all his own eggs, and after the broiler season is over, with him about the last of May, he markets his eggs. Each week he ships all he receives from his hens, and, up again to hatching time, he has a regular weekly income. If all engaging in this business would follow Mr. Reed's plan there would be less reason for closing up shop. He keeps as many as two hundred hens.

"I am undecided what to do," writes a FARM-POULTRY reader. "After reading editor Hunter's egg records I sort of have an egg farm fever; and then reading your broiler articles am a little broiler inclined. Which will pay me best? I fully understand that I must start small, invest capital, and all that. All I want to know is, which is the best branch for me to follow." Both. No broiler farm should be run without an egg farm attached to it; and a broiler plant would turn eggs to profitable account when the market rate for them is low. Profit by the experience of Mr. Reed, as given above.

"What plan of fattening is adopted by the Hammonton broiler raisers, and do they use fattening coops?" asks a Pennsylvanian. The only one who uses a fattening coop in Hammonton, is George W. Pressey. The floor of this coop is slatted, so that the droppings fall on a movable platform beneath. Every morning this platform is cleaned off. In a week, generally, the desired weight can be secured, and the birds go plump to market. Scalded corn meal, boiled potatoes, meat scraps, etc., are in the bill of fare for fattening. The other broiler raisers of the town use the same fattening feed, but keep the birds in the brooders. I rather favor Mr. Pressey's plan, as it costs less time and expense, and the birds are in a better condition.

"Will you please suggest to me a good plan of furnishing shade to my chick yards? The sun lays very heavy there, and my late hatched chicks suffer considerably." I have grape vines planted at end of my brooder yards. They give a nice shade by training the vines over the top. Besides giving the shade, I have plenty of fruit from the vines.

"How long can I keep eggs for hatching?" The fresher the eggs the better. However, they hatch well at three weeks of age, and I have kept them four weeks with good results. Eggs for hatching should be kept in an egg crate in a cool place. Place the eggs on end, and turn every other day.

"Is a clay soil advisable for a broiler farm?" No, indeed. One of the requisites of hardy and thrifty stock is a dry location. Dampness is fatal to all stock. A sandy soil is best for a broiler farm.

"About how do the weights run in broilers according to the seasons?" In February, one pound each; March, one and a quarter pounds each; April and May, one and a half pounds each; after that they are allowed two pounds each.

"Are any automatic attachments ever put to brooders?" No; and it is a good thing there are not. No brooder will do good work without careful attention, and no automatic appliances could be invented that would take the place of brains and good common sense.

"Can you tell me why so many have failed in Hammonton, and why so few are in the business now? Is the market any worse than six years ago?" Some years ago, there was an "over-booming," as editor Hunter expressed it in FARM-POULTRY, and every fellow that was out of a job, or looking about for a "big bonanza," and had a little cash to invest, took the fever. As many as forty brooding houses sprung up at one time in the town; but gradually the romance of the affair died out; and, again as Mr. Hunter puts it, "the reaction which inevitably follows inflation," gradually dwindled down the number, until today not more than a dozen houses might be termed in working order. But it must be said to the credit of the remaining ones, that they did remarkably well, and are much encouraged. Nearly all of them began small, and gradually built up their business as their experience and cash allowed. The market today, if anything, is better than it was six years ago.

CHAPTER X.

MR. JACOBS AND HIS OPINIONS.— RICHARD H. WHITE'S METHODS.

Mr. Jacobs, in the September issue of the *Poultry Keeper*, in his reply to the "Impression" article by the editor of FARM-POULTRY, gives this excuse for not combining an egg farm with a broiler establishment:

"It is true that Mr. Hunter suggests that the hens be kept to lay the eggs. It is also true that some do so; but it takes a great many hens to provide eggs for half a dozen three-hundred-egg incubators every three weeks, and buying is a necessity, as hens do not always lay when you so desire, in the winter season."

Now Mr. Jacobs is not definite when he says "it takes a great many hens to provide the eggs," etc. What does he call "a great many?" Let us make a small calculation, and see just how many it would take. It would be a very modest estimate to suppose that on an egg farm each hen would average one and three-quarters eggs a week, or ninety-one eggs each per year. We will say that ninety-one eggs are all we can get. Three hundred hens will lay (at the rate of one and three-fourths a week each) five hundred and twenty-five eggs per week, or fifteen hundred and seventy-five eggs in three weeks. Not quite enough eggs to fill six three-hundred machines — falling two hundred and twenty-five short of the amount — but still enough hens to lay the eighteen hundred eggs required, if given good care. Without fear of being extravagant, I can say that that number of hens will keep six incubators busy.

Mr. Jacobs says some *do* raise their own eggs. So they do. Mr. Seely told me that he owed all his success to his hens, who supplied him with plenty of good eggs. Mr. Alfred Reed, who also runs a large house in Hammonton, told me the other day that he never buys an egg. It pays him best to raise them. I cannot say just now how many hens Mr. Reed keeps, nor the amount of broilers he turns out in a season, but intend to have the full information for my next article. In the meantime I cannot see that there has yet been any argument brought out to prove that it is not the best policy to combine egg raising and broiler farming. Now what surprises me most is, that Mr. Jacobs, a man who has been studying and experimenting on this subject for so many years, should still stick to the "no-egg-farm" idea, when right in the town of Hammonton so much trouble has been experienced on that score.

But there are two sides to this question, and Mr. Jacobs has the right to claim the very exception which I met this morning. I called at the broiler farm of R. G. White (whose interview I give in this article), and found he was unusually successful at buying his eggs. In short, he has contracts with a number of farmers who keep grades, and whose flocks are small. He knows exactly what to expect from the eggs he buys, as they are from hens that are properly mated and composed of the right blood. I say, where a man can buy enough eggs from just such parties, the egg farm might be another question. But how many can do that? Mr. White's case is the only exception I know of in Hammonton. Just the other day a broiler man called upon me and wanted to make an arrangement to buy all the eggs I could raise. "How about those black fowls?" said I, pointing to a pen of Minorcas. "I don't care," he replied, "what color they are — I must have eggs, and it matters not what fowls lay them. I get just as much for white skinned birds as I do for those that have yellow skin." He said that eggs were getting scarcer every year, and that if he could not meet with better success he would have to quit business, or buy up eggs in the cities. That is but a single example of the difficulty of obtaining eggs, and, to my mind, a good argument in favor of supplying your own demand.

Mr. Jacobs, in the same issue of *Poultry Keeper*, thinks my estimate of fifteen cents per pound (which includes the cost of eggs, feed, brooding, etc.) " is correct, so far as the cost of a pound of chick is concerned;" "but," he adds, " the cost of the food only to produce one pound of flesh (chick) seldom exceeds five cents. This fact has been demonstrated by actual experiment in weighing the food and also weighing the chick, at successive stages of growth." Very well, I will then allow that Mr. Jacobs is correct on the price of food, as he says my estimate of the ultimate cost is right — fifteen cents; so the whole argument dwindles down to this: Mr. Jacobs can buy his feed cheaper than I can, while I know how to get the eggs and oil, etc., at a cheaper rate than he does. The result is practically the same, and my estimate of fifteen cents per pound to bring the chick up to a marketable size, stands.

As I have already mentioned in this article, this morning I made a visit to the broiler farm of Richard G. White, on Fairview street, and when I left it I felt that I had a good share of practical knowledge for FARM-POULTRY readers. Mr. White is a plain, hard working man. He is in the fruit business in summer, and the broiler business in winter, and he speaks very encouragingly of his occupation. He has no secrets, and does not flinch one bit in answering any and all queries that may be asked him. Last year he had a capacity for one thousand chicks, but this year he has increased to three thousand.

His system of brooding last year was the Packard bottom heat method; but this year he will use both the top and bottom heat, to fully satisfy himself which is best. Mr. White has a novel way of testing the heat in his brooders. He places a thermometer on a stand about two and a half inches from the floor, as he thinks this gives the best average temperature. He begins the heat at eighty degrees. This is the lowest temperature I have yet seen in starting young chicks, but the appearance of Mr. White's chicks would indicate that it is not a bit too low. There is more danger in getting the heat too high. One writer in particular uses one hundred degrees; but as I have never seen his chicks, I am unable to say if they do well at such a heat. However, Mr. White's product gives living testimony to a lower temperature. Mr. White uses the Pressey hot air incubators, and is very well pleased with the results; but he admits that he has a weakness for hot water machines, as a past experience in that line convinces him that hot water gives plumper and better chicks. He will try both methods this year, and the one which suits his purpose best will in the future be used.

In the matter of feeding, Mr. White is also different from the rest of the broiler men of Hammonton. He makes a regular johnny-cake, but leaves out the vinegar and soda which a number of writers suggest, and this cake he feeds the youngsters until they are one week old. It may also be stated that he puts animal meal in the cake instead of the usual prepared meat. He likes the meal for several reasons, but principally on account of its fineness — the chicks are bound to eat it. After the first week he drops the johnny-cake, and feeds a mash composed of equal parts of corn meal, bran, and middlings, with the usual amount of meat scraps. Now as at this stage of the chicks' growth bowel trouble is apt to show itself, he keeps a close watch, and regulates it with middlings — lessening the quantity of middlings if the chicks become costive, and increasing the amount if they have looseness of the bowels. After trying this method for several years, he says he finds no trouble in keeping the chicks in the right condition. He feeds wheat and cracked corn only as a relish. Several times during the day, between meals, he throws several handfuls of the grain to them, thus getting them to exercise in scratching and running about. Grit is constantly kept before them. He uses small sharp gravel, and likes it better than oyster shells for the purpose. When the young chicks are put in the brooder a pan of ground charcoal is placed before them. After several days if it is noticed that they do not eat any of it, charcoal is put in their cake, and afterward in their mashes. "They must have charcoal," said Mr. White, "and I believe my success in raising chickens is principally due to the charcoal." But Mr. White does not stop here. He scatters oyster shell lime on the brooders every morning after he cleans up. They pick at it with a relish — and a pan of sifted coal ashes is put fresh before them every morning. It serves a double purpose: both as a dust bath, and a treat in various substances which they pick out of the ashes.

"Did you have good success last year in raising chickens?" I asked Mr. White.

"Very good," he replied. "My capacity was small, but out of fifteen hundred that I hatched, I lost only fifty; the remainder I marketed, which my books will show," and Mr. White made an effort to get his books to verify his statement, but we assured him that was not necessary, as appearances at his place were enough to indicate prosperity.

"Why do you not raise your own eggs?" I asked.

"Well, I have two reasons. First, I do not have much faith in **egg farms** unless fowls can have a free range. I have not ground enough to keep fowls on such a plan; so I do the next best thing by buying up my eggs from small flocks, and from birds that have my personal supervision. That is, they are mated and composed of such bloods as will give good broilers."

I tried to show him that the hens would be profitable in limited ranges if properly fed and taken care of; but Mr. White's opinion did not coincide with mine, and as he had such an excellent arrangement to get eggs, I did not use any further argument on that score.

The size of the "mother" (brooder) in Mr. White's pens is three feet square; the brooder floor is four feet by four feet six inches; the yard inside the pen is four feet six inches by five feet; and the outside yard is four feet six inches by sixteen feet. One of these pens he reserves for a hospital in which are put cripples, a dozen or more of which are apt to show themselves every season. The house of last year contained ten pens, and in the nine of them (one being counted out for the aforesaid hospital) Mr. White was compelled to crowd fifteen hundred chicks.

"Was not that a dangerous move?" I asked.

"Decidedly so," he replied, "but I could not help it. I had contracted for the eggs, had to take them, and, of course, was compelled to put them in the incubators. The eggs were remarkably fertile, and my incubators never worked better. Out came the chicks, and I had to care for them."

Considering the crowded condition of Mr. White's house last year, and the remarkable success he had in raising his chicks, I deem his method of feeding worthy of attention. He will make one change this year, however, and that will be in the feed the first week. Instead of the johnny-cake he will make a cake from the recipe used by Mr. Howe, which I have given in a former article in FARM-POULTRY — only, instead of the meat scraps he will use animal meal, as he likes that better for young chicks.

"What do you consider the proper number of chicks for your brooders, to avoid crowding?" I asked.

"I always aim," he said, "to have not over one hundred and fifty in the brooder the first week, although two hundred can be managed. When they run to three-quarters of a pound I thin down to one hundred. After that as a chance offers, I take a few out, until only seventy-five are in at the time they reach the marketable size."

"Have you ever tried summer hatching?" I asked.

"About two years ago, I gave it a trial, and met with very good success. never getting less than twenty-five cents a pound, and, in most cases, thirty cents. As there was not much cost in raising them, I felt highly encouraged in the results. It was my intention to keep on hatching last summer, but putting up the new buildings and other work prevented me from doing so, but I shall keep right ahead from now on."

"Well, now, Mr. White," I concluded, "what is your candid opinion of the chicken business in all its branches, compared with other farm work?"

"I have tried farming in all its branches, and have been fairly successful, but for capital invested, and the amount of real solid work, I can make more money out of the chicken business than I can out of any and all the other branches of farm life. I am not an enthusiast, but practical results tell — and I find broiler raising the most profitable of all the branches in the chicken business."

Thus ended an interview that I consider worth considerable to my readers, and as it comes from the lips of one of the most practical and successful men I have ever had the pleasure to meet, I unhesitatingly report it, and believe it will bear good fruits.

It would not be fair should I close this article without saying a good word for Mr. White's help. Good help in the chicken business is as scarce as hens' teeth, but Mr. White has the cream. It consists of his wife and daughter. Go to his place when you will, and you will find the full force hard at work. Every movement seems to be closely watched, and nothing seems too much trouble for the comfort of the little orphans. That the work pays them, is evidenced by the interest they take in all the operations. Mr. White has had some very flattering offers made him to take charge of a large concern in the northern part of New York, but he would rather run his own place and be at home.

A lady residing in Decorah, Iowa, writes: "Will you please tell me, through the next issue of FARM-POULTRY, what effect it will have on the chickens hatched, if we do not have enough roosters with the hens? We have hatched quite a number of chickens, but they take the dysentery, and dwindle away and die. What is good for the disease named? I have raised chickens a good many years, but never before had such luck with them. Have raised over fifty this spring without mothers, and had better luck with them than those that had mothers."

The dysentery was caused by something in their food that did not agree with them. Again, the trouble may be due to a cold. I would advise a feed as given above, in the interview of Mr. White, using the middlings as the regulator. If the eggs showed sufficient fertility, there were plenty of males in the flocks.

"Do the broiler men ever use Leghorns pure, for broilers?" asks Mr. Colgan, of New York. Not as a rule. They prefer them crossed upon another breed, as for instance Plymouth Rock, or Wyandotte, or Brahma.

CHAPTER XI.

THE VISITORS.—ALBERT REED'S BROILER AND EGG FARM.—COMBINATIONS.—COLO-NIZING BROILERS.

I see the question, "What shall we do with the visitors?" is perplexing FARM-POULTRY readers. It makes the Hammonton broiler men laugh. When Hammonton saw that the visitors were about at the end of their rope, they tacked up the sign, "No Admittance," and the poultry world became shocked! Shocked because Hammonton wanted fair play! Shocked because her poultrymen could not see the benefit in entertaining as many as forty visitors a day, and teach each and every one the mysteries of poultrydom. Ever since those notices appeared every failure in Hammonton has been made a big fuss of; but she would rather receive the cold shoulder from the public, and be let alone, than have the public idolizing her, and keeping her from work. We must certainly say "quits" to the visitor unless the visitor will take a hint and not ask the earth of a man who is willing to at least be obliging.

A reader of FARM-POULTRY, Dr. E. Mussina, of Austin, Texas, gives a good idea, and one which might be of advantage. He says:

"I have been reading your articles on the broiler business in FARM-POUL-TRY, and find your exposition of the whole affair to be sound, and it exactly represents the deductions from my own amateur experience through several years. In the August number of FARM-POULTRY you say: 'Every now and then applications come in by mail for jobs on a broiler farm, and equally as many letters are sent me asking for an experienced man, and I cannot help either,' etc. You give the advice: 'Apprentice yourself to some practical man, and then let the employer give the recommendation.' This is the only practical plan I have been able to see myself, but such is the incompetency for learning, as found in the average man, I presume many proprietors in the business will not be bothered even for a money consideration for their trouble. I presume you will agree with me on this point. Now, in view of the total inadaptability of so large a percentage of those who choose to seek jobs in this line, would it not be a good plan to suggest to some one to advertise to receive applicants for a stated fee, pass upon their fitness after an examination of their adaptability, and receive only those who could pass on reasonable qualifications, get them a boarding place near by, and give them for a consideration the privilege of following the manager into the incubator cellar or rooms, and on his daily rounds, in order to observe all the details of the business, and receive the manager's instructions in the shape of a lecture? This for a stipulated time of one to three months or longer, as seems best. At the end of one or three months, rent (or sell) to the student an incubator or brooder, which can be stocked by proprietor for value received, and run for two or three hatchings at student's own expense and risk, the hatch, if successful, to be the property of the student as an incentive. The student can then work for some one in a position of small exaction under recommendation of proprietor, his preceptor. It seems to me this might be worked with remuneration to proprietor and satisfaction to student."

This letter, which I quote at length, contains some points which it might be well to consider. If each of the large concerns would take one or two apprentices in this way the broiler business might be considerably benefited. Arrangements could be made by which the broiler house should receive so much cash and so much labor to pay for the time lost in teaching the student. If we would turn out broiler men in that way, we would have fewer failures. Probably some of the readers might advance some ideas, and FARM-POULTRY could devise means by which we could get up a system of instruction that would benefit the market poultry business.

I made a visit to the broiler farm of Albert Reed this afternoon, and found that gentleman as busy as a beaver, but ready to answer all the questions I could put to him. Mr. Reed is not a bit vain; in fact, he thinks he has much to learn yet, but those who have ever witnessed his success are apt to think that he is far from being an amateur. Mr. Reed has the regular style of brooding house, containing twenty pens. The calculation is to put one hundred in a pen; but Mr. Reed says he never puts in more than seventy-five, and thinks that number more than is comfortable. "Fifty chickens crowd each pen plenty," he said. The heating system is the top heat method. Formerly the bottom heat plan was used, but last season top heat was introduced with a Bramhall heater.

"Which method do you like the best?" I asked.

"In my experience," he replied, "I see no difference. If anything, I like the bottom heat plan the best, for the reason that I can clean the brooders more easily than with the top heat, on account of the pipes interfering; but as I now have the top heat in, I will let it go at that."

Four Hammonton (Pressey) incubators of three hundred capacity, are charged with eggs, which will turn out broilers to be marketed about February.

In the matter of feed, Mr. Reed says they tried nearly every known method, and found no material difference, as long as all other care was equal. Some lots he started with cracked wheat; others with rolled oats. He saw no difference in that. He, however, most generally starts with cracked wheat, and follows after the first week with a cake made of corn meal, bran, and prepared meat. Of course the usual side dishes of oyster shell, charcoal, etc., are not forgotten.

As stated in my last article, Mr. Reed raises his own eggs. At present he has between one hundred and eighty and two hundred hens, and, by careful feeding and management he secures as high as *one hundred and twenty-five and one hundred and thirty eggs a day in winter*, which is *more than enough to supply his incubators.* "I am often compelled to sell eggs," said Mr. Reed, "to keep them from spoiling." This I thought another point for the argument that *it is* possible for enough hens to be kept to furnish the incubators. Mr. Reed is positive that his hens would fill five incubators, but four machines are enough to keep the house going. All the hens on the place are dunghills, mated up to good males; but I did not see a hen on his place, notwithstanding they are all common, that did not show the best type for laying qualities. Those who have ever had any experience with egg farming can readily distinguish the profitable from the unprofitable hens—and I must say Mr. Reed's selections are very good. Another point worth mentioning is, that no hens are kept over two years. I noticed considerable Jersey Blue blood among the hens, and Mr. Reed counts them his best layers. The rest show more or less Leghorn blood, although none of them are two-thirds of any breed. I saw no trace of Asiatic blood, but there are many brown egg layers in the flock.

"How do you feed your hens?" I asked.

"I scald bran and shorts, (middlings), and add sour milk or whey. About twice a week I put in the mess some prepared meat scraps. This makes their breakfasts. At night I give them wheat or oats, mostly wheat."

"You feed no corn, do you?"

"No, it is too fattening. If we have very cold weather I occasionally give some corn at night, to keep the hens warm, but I am very sparing with the grain, as I do not like it for laying hens."

Mr. Reed conducts a general farm. He raises vegetables and fruits, making a specialty of corn and white and sweet potatoes in summer, and what eggs his hens lay; and broilers and eggs in winter. He also keeps a horse and a cow. The latter, he thinks, belong to a poultry farm.

"You evidently do not believe in hatching broilers in summer?" I inquired.

"Oh, yes, I do; but I have not time to attend to them properly during the summer season. If I hatched then, I could only half attend to them, and then it would not pay. I find just as much profit by running the farm in summer, and at the same time it gives me a rest from the care of the broilers."

"What do you think of the broiler business, as a business?" I ventured.

Mr. Reed smiled as he said: "Well, I think it is mighty steady work. It is work that requires close attention from early till late. It won't do for any tired fellow. And it won't do to hire help, for no one will take the same interest in it that you would yourself. As for profit, I know it pays me; that is, I make good wages at it for winter work."

Mr. Reed brought out the point that broiler raising was a good adjunct to the farm. It is an item which some of our farmers, who are slaving themselves during the summer to live in winter, might well consider. What he is doing could be done on every farm, and in a few years it would be seen that the winter's work was as profitable, if not more so than the labors of summer. Poultry in any of its branches makes a good adjunct to the farm.

"What branch of poultry culture do you like the best?" I asked.

"If I had to commence over again," he said, "I would run an egg farm, and have a small brooding house to use up the eggs when the market price was down. I believe, taking everything into consideration, that egg raising pays the best, and a brooding house to help out when eggs were cheap, would add still more to the profit."

Mr. Reed does all his own work. He says that he has his hands full during the entire winter running his incubators, twenty brooders, and two hundred hens; that he has reached the full extent, and could not possibly handle any more, and does not believe it would pay him to run on any larger scale.

Visiting places like Mr. Reed's gives one good material to work upon. He is a hard worker. He began small, reads, experiments, and is always glad for advice. He does not claim to be an expert, yet his success is enough to make some of the so-called experts blush. He never boasts. What he tells you he is ready to prove, and while not growing rich, he makes annually more money than some of the gigantic affairs we read about, but which seldom hold out.

The question which Hammonton is trying to solve, and which, no doubt, other sections are debating, is whether it will pay them to quit the fruit and general farm work, and hatch broilers the entire year; or whether it will pay best to continue in the old way. Certainly, at the prices experienced the past year or two, summer broilers pay. Will it hold out? Some claim that it will, while others are afraid that too many may undertake it, and the bottom will drop out of good prices. There is no danger of the market ever being glutted in winter. Again, others are considering the advisability of combining egg and broiler farming. Will such a combination pay better than broilers and fruit—or, broilers and vegetables? A farmer in conversation with the writer said that he proposed running a combination of broilers, eggs, lima beans, onions, and sweet potatoes. He said there was money in each and every one of them. He further said, that of all farm crops, he could make more money out of the three mentioned vegetables than all the rest combined, and he proposed to drop all but those, and take up broilers and eggs to fill up the time. These are all good suggestions, and will stand trial. I have always been opposed to running risks on one article. I opposed broiler farming as an exclusive business, as I could see no means of salvation if the crop should be a failure. A combination would help out. "One hand would wash another," figuratively speaking.

A correspondent of FARM-POULTRY would like to know which method of brooding is the best: One large house, or several small houses on the colonizing plan. Undoubtedly the colonizing plan is the best. Mr. Pressey, of this place, gave that method a good test. He had a separate house for each brood, and I never saw better chicks in my life. It gives more labor than the one house plan, but it is a question if it is not more profitable, despite the extra labor. In case of sickness there is no danger of disease spreading, and if the houses are set some distance apart, no fences are needed, and each flock can have an unlimited range. It comes nearer the hen fashion. In fact, it is this unlimited range that makes hen-brooded chicks so much more rugged than those confined to brooding houses, and the work necessary to run a number of such flocks will not be much more than that many hens, with the advantage that the flocks can number a hundred instead of a dozen, as it would be with a hen. I believe that the business will eventually drift to the colonizing plan. Another argument in favor of it, is that there will be no danger of heavy loss in case of heat going down in the brooder. One house in Hammonton one year lost several hundred chicks by bowel disease, caused from the heat going down, due to a breakage. The colonizing plan can also be made better rat proof, and it can in many other ways be made the safest method for brooding, and I would advise it for the beginner especially. Still another point in favor of colonizing: The houses can be so constructed that they can be readily moved about. By moving them each year to new ground, there will be less sickness, and at the same time the ground is enriched for planting a little truck, or seeding to rye for the chicks.

CHAPTER XII.

GEO. W. PRESSEY'S PLAN FOR RAISING BROILERS. — HARRY PHILIPS' GIGANTIC ESTABLISHMENT. — VALUABLE POINTERS. — GOV. LEVI P. MORTON'S PLAN. — BROILERS AND DUCKLINGS. — HOW TO GET FERTILE EGGS.

George W. Pressey, of Hammonton, N. J., has done considerable towards making a success of the colonizing plan, and without a doubt is the father of the idea for broilers. Mr. Pressey has disposed of his interests in the incubator and brooder business, and as his daughters (who always took charge of the brooding work) have too many other cares to attend to the broiler work, it may be said Mr. Pressey is entirely out of the field for the present. I have repeatedly visited Mr. Pressey's place, and undoubtedly the finest looking chicks I ever saw were in his brooders. He was a great believer in fresh air, and his chicks always had the appearance of being benefited by the "exposure." He had erected houses about four feet square, and probably four feet high, with a slant roof. They were well made, and the roof shingled. In each one of these he had placed one of his brooders. The floor of the building was made of boards, underneath which was stretched one-half inch wire netting. This prevented the rats from getting on the inside. The fronts of these little houses were covered with cloth. Mr. Pressey said: "We raised a hundred chickens until ready for market, in each brooder. When the weather was not too cold I removed the brooder cover the last two weeks, and placed some roosts in the back of the shed, two inches above the floor and brooder, for the chickens to roost upon, placing a piece of board a foot square on the roost over the warm air pipe to prevent the dirt from falling down the pipe, and which also gave me a chance to spread warm air on cold nights. The yards and brooders were always kept clean."

"Is not the colonizing plan apt to be objected to on account of its much use of fresh air?"

"Probably some will think that way. I never was afraid of snow or cold. Chicks a week old will stand more cold than you can, provided they have pure air to breathe, and a warm brooder to run under when they feel chilled."

"What temperature did you keep the brooder?"

"The floor was between fifty and sixty degrees, and the warm air under the hover from seventy to ninety degrees, according to the age of the chicks. It will not do to keep the brooder too hot."

A brooder house was built later, when it was found that the houses erected on the colonizing system could not hold all the chicks hatched. It is a regular shed, fifty feet long, and ten feet wide. It is divided into ten sections of five feet each. Each section has one of his brooders in it, in the back left-hand corner. The space from the brooder to the side of the section is floored over even with the top of the brooder, and an inclined platform is placed in front to enable the chicks to run up and down from the ground to the brooder. Each partition has a wire gate in the middle, through which the attendant passes in feeding. From the gate backward the partition is of boards, and from the gate forward of wire netting. The posts are two feet high; rafters seven feet ten inches long; studding for gates six feet, all 2 x 3 inches. The sills and back plate 2 x 4 inches. No front plate is used, as it would be in the way. The front posts are held in place by boards one foot apart, running across the shed and forming the bottom of each partition. A short brace is also used on each side from sill to post. The front of the shed is closed by a wire screen 2 x 5 feet, and this screen covered on the outside with a cloth curtain (hanging from top of the screen) to be used when the weather is very cold or stormy. The back and ends of the shed are made tight; with the back roof and one-third of the front roof boarded and shingled. The lower two-thirds of the front roof are fitted with sash, and covered with cloth or set with glass (Mr. Pressey prefers cloth), so that they may be opened on every pleasant day. The front screen is made so that it may be removed, and thus allow the chicks the 5 x 16 foot yard in front. An opening through the incline in front of the brooder allows the lamp stove to be drawn out to refill. All the labor and perplexity of keeping a coal fire night and day is saved, and the expense reduced from twenty-five or thirty cents for each brooder per week for coal, to about nine or ten cents for oil. This makes quite an item in the expense account. One good argument against the large house plan, where the hot water pipe system is used is, that it requires the same amount of fuel to heat up for one flock as it would if the entire house was filled. In the colonizing plan the expense is attached to the actual number in the brooders, and there is no waste of heat.

Mr. Pressey's aforesaid shed must not be confounded with the regular styled brooding house, for it is just the opposite, and made to come as near the original plan of colonizing as possible, Mr. Pressey living on a few lots, and not having the ground room for more separate brooders. He uses no glass at all in his building, finding that cloth does far better. Having each brooder heated by an oil stove, he has no waste of heat.

"In this shed the last season I ran," said Mr. Pressey, "we raised nearly five thousand chickens to the age of four and five weeks, and found it to be a perfect paradise for chickens, the largest weighing at the end of seven weeks one pound and seven and one-half ounces."

Harry M. Philips, the proprietor of the largest brooding house in Hammonton, says that if he had to do it over again, he would have none other than the colonizing plan,—that chicks will grow better and remain healthier than when placed under one roof. He has tried both methods.

Speaking of Mr. Philips, reminds me of some of his business principles. The fact that he will not admit visitors, has created the impression that all the show of his plant was in the outside appearance; but this is wrong. Mr. Philips is one of the hardest workers in the business, and he has his work down to a science. All his stock when shipped to market is labeled, and as he is very careful to send none but gilt edged goods, he generally receives a few cents more on his shipments than is given the average lot. This same plan is adopted by the Long Island duck raisers, the latter having a tag fastened to the neck of each duckling shipped, showing where it is from, and what can be expected by the purchaser. Mr. Philips' incubator room is a model of neatness, and so constructed that it means an even temperature. The room is lathed and plastered, and kept at a certain temperature the entire time of hatching. Instead of running his brooding house the entire winter and spring, Mr. Philips fills up all his incubators at once, and has one or two grand hatches, filling the entire house, which will hold five thousand chicks at one time. In this way much is saved, and the work is accomplished in one-fourth the time. The eggs are cheaper in spring, and less fuel needed for brooding. Besides, he can market the lot in two or three shipments. He says, that notwithstanding the drop in price by the time he gets his broilers in market, he makes just as much as when they are out in the cream of prices, as his stock is produced at a saving of labor, fuel, etc.

There is no question about it, the business of broiler raising can be gotten down to much finer points than we have reached yet, and the question of marketing is yet in its infancy. Success in this will depend first, on the quality of goods marketed; second, properly advertising the same; and third, the proper customer. When the broiler raiser raises his own eggs, and keeps a pure breed or a systematic cross for furnishing those eggs, he will have better goods to offer. Take for instance the cross I mentioned in a former article :—Houdan cockerels crossed on Cochin or Brahma hens, or the Wyandotte in its purity; and there would be a uniform lot of birds in weight, plumpness, and yellow skin. Now if such a lot of birds are proberly labeled as coming from the farm of Mr. So-and-So, is it not reasonable to suppose that this man's goods will have a regular line of custom? They will be sought after. On the other hand, as a rule, there are a variety of shapes and conditions in the broilers marketed from individual farms. Another point for the enterprising man who has attractive stock to sell, is to seek as much retail custom as possible. If the club houses and large hotels knew they could buy gilt edged birds from the farm direct they would be only too glad to give their standing orders, and the commissions of middlemen would go into the hands of the raiser.

I have been asked what is the meaning of "Philadelphia broilers," and if there are large brooding houses in the vicinity of that city which gave it that name. The term is applied by the commission men to all broilers that are shipped from New Jersey, Pennsylvania, and other neighboring states. It is used more as a classification.

As it will be remembered, I have always held that broiler raising should be an adjunct to some other business, and not an exclusive business. Some combine it with fruit, and that is the general plan adopted in Hammonton. Others use it with egg raising, which makes an excellent combination; but to my mind a unity of the three would be better—fruit, broilers and eggs. Up in Long Island they combine broiler raising with duck culture, and it is an excellent idea. As the ducks can only furnish eggs during the months from February to September, as a rule, the broilers could be turned out from September to February, and at a season when they would be bringing the best prices. Thus the incubators would be run the entire year, and at a profit.

A correspondent from Columbus, Ohio, would like to know what are the requisites of a good hatch. He acknowledges that the man is the most important; but in this case he supposes the man to be equal to the task. I should say that, first, a good place for the incubator; second, a reliable thermometer; third, a safe burner. As to the proper place for an incubator, I have gone to some extent in one of my former articles. The importance of a reliable thermometer and safe burner cannot be too strongly given. Unless a thermometer is true there will be no guaranty that the proper heat is given. Thermometers with the mercury parted should never be used. Likewise much depends upon the burner. If a poor one is used, there is great danger of the wick working loose and increasing the heat when you are not about. A steady flame is of great importance. Wicks should never be trimmed with scissors. The best way is to scrape off the hard crust with a burnt match. G. A. McFetridge, one of the most successful operators of incubators in this country, cuts his wick with a knife. He turns down the wick a little, just enough to not expose any of the sound part above the burner, and then placing the heel of the knife on a level with the top of the tube, he draws the full length of the blade while crossing the tube. This makes a clean cut. But all this trouble could be saved by using the Sunlight Porous Carbon Wick, for sale by S. G. Robinson, 29 Purchase street, Boston, Mass. I know by experience that this wick will give a steady flame, never needs trimming, and emits no odor nor smoke. Besides, a single wick has been known to have been burned every night for two years, and this without being trimmed, or losing its original brilliancy. I mention these facts as I deem them valuable to all running lamps, and who need a very steady flame.

The matter of green food is often a perplexing problem to the broiler raiser, but if he adopts the colonizing plan he will have very little trouble on that score. It must not be forgotten that my idea of this scheme is to have the little brooding houses scattered, and no yards attached. Each flock is to have its freedom. All the ground should be plowed up, and rye sown. In a few weeks it will be up to such a height that the little chicks will have a feast. As it is not killed by winter, it furnishes a splendid crop for them. Early in spring it shoots up again, and the chicks will have a harvest until the rye has grown too large for them, when it can be mown down or plowed under, and wheat sown. This will serve them well until that, too, has grown too large, when oats can follow. Thus a constant growth of green food will always be within reach. By way of variety chopped lettuce and onion tops may be given them, which they will enjoy with a relish.

When the writer one day asked James H. Seely, who was then superintendent of the broiler department of Governor Morton, Rhinecliff, N. Y., how he fed his chick, he replied:

"We are like most people," he replied, "feed anything that is good, such as corn, oats, wheat, cotton seed meal, ground meat, potatoes, cabbage,— all of these in a variety of ways — so as to keep the appetite good. Cotton seed meal is not used much, but I find it good when given with judgment. I use a great deal of skim milk; they have it to drink all of the time. I get it within an hour after it is milked, and while yet warm. The chicks are very fond of it, and I think it will pay better than to feed it to pigs, which bring six or seven cents per pound, while chicks bring twenty-five cents to forty cents. I use a little plaster on and under the brooders; it makes them smell sweet. I also use considerable hay seed and chaff in the pens, as it keeps the chickens busy; besides they eat leaves and seed, which does them no harm."

Mr. Seely believes in starting the chicks with a temperature in the brooders of from ninety-five to one hundred degrees. He does not think a broiler can be gotten up to two pounds before about twelve weeks; but at present the best markets want lighter weights than two pounds. Mr. Seely does not find much difference between the hot air and hot water incubators, excepting that the hot air machines will last longer, and if it happens that the heat should go down, it doesn't take so long to get it up in a hot air machine as in a hot water make.

A man can be in the poultry business exclusively, and still have a combination. A broiler and egg farm is a good idea — sell the eggs when prices are high, and turn them into broilers when prices decline; but broilers and ducklings seem to take better among the broiler men.

The incubators could start on duck eggs the latter part of February or the beginning of March, and be kept going right along until about the first of September. During the time the duck eggs are being hatched the hens' eggs can be marketed. Of course that part will come at a time of the year when prices for eggs are not at their best; but, nevertheless, taking the good with the bad, there will be sufficient income from the sale to more than pay the expense of feeding the hens during the summer. Then when the ducks cease laying will be a good time to hatch broilers, which will give broilers in January and February, the beginning of the high prices for that article — and as the ducks will not begin laying until the latter part of February or March, the entire fall and winter can be turned to account in broiler raising.

This plan, I understand, is somewhat similar to the one worked by Mr. Rankin, of South Easton, Mass., only that Mr. Rankin turns his chickens into roasting fowls instead of broilers. Which of the two would prove most profitable, depends mostly upon the markets you wish to serve.

Probably a few hints may be in order. To have fertile eggs during the fall and winter it is necessary, first, to have good vigorous stock; second, good feed; third, warm houses; fourth, sufficient male birds; fifth, young stock; and sixth, plenty of exercise. By "good vigorous stock," I mean hardy birds, of a quick growing nature, which have the desirable qualities of plumpness and meat supply.

Next we come to good feed, and on that subject alone probably more articles have been written than on any other. It is not my intention to give a bill of fare. The reader can secure valuable information on that point in our book, "Profitable Poultry Farming," which the publishers will send for twenty-five cents. But I do wish to add that it must not be forgotten that fattening foods will *not* make eggs, neither can an overfat hen lay fertile eggs. Green food in winter must not be forgotten, and in order to best supply it, it will pay poultrymen to have hot-beds, in which can be raised right through the winter all the lettuce (which, by the way, is the best green food) we can give old or young stock. No matter what bill of fare is adopted, there must always be a liberal supply of sharp grit on hand, for without the proper grit the fowls will be unable to thoroughly masticate their food. This is a thing that must not be overlooked. The advice to have warm houses explains itself, and it means all it implies. I do not advocate artificial heat in the hennery, but I do believe in good roofs and thick walls. The animal heat will do the rest. In a house well protected so that the cold cannot come through the heavily lined walls, I had a flock of Black Minorcas laying all winter, and not a frozen comb in the lot.

Do not allow the manure to accumulate, with the idea that it will add warmth. There is more to be gained by clean quarters than by the extra heat that the manure may give; besides, lice can live all through the hardest winter if you give them "good warm manure" to nestle in. Clean out the manure, and with it you are cleaning out the lice.

What do I mean by sufficient male birds? I have adopted a plan which works admirably. It is not original with me, however; but as I cannot recall who first wrote it, I do not now know to whom to give the credit. I have three male birds for every two yards. The birds are numbered one, two and three, and the pens one and two. I begin by putting cock No. 1 in pen No. 1; cock No. 2 in pen No. 2, and cock No. 3 in a separate pen alone. Each day they are changed; that is, at night I take cock No. 1 and put him in pen No. 2; and cock No. 2 I put in the resting pen where cock No. 3 now is, and bring cock No. 3 over to pen No. 1. In that way, each night, the changes are made. One day rest out of three certainly does put life into the male birds, which gives me the double advantage of keeping more hens in a flock than I would dare do if the same cock was in the yard the entire season. By young stock, I wish to imply birds not over two years of age. Of course, hatch out the pullets each year during the months of April and May, for fall and winter layers. I do that, and keep in addition the pullets of last year for breeders. I find that the eggs hatch better from the two year old birds; or rather they give better offspring, so the first year I do not mate up the pullets, but market those eggs, and the following year I introduce cockerels in the pens for breeding purposes, market the two year olds for prime roasters, and again send the pullet eggs to market. This gives me the chance of getting some of the good prices obtained for eggs and roasting fowls at that time of the year, and at the same time does not interfere with my broiler supply.

Now, last (but as important as the feed the hens get), comes exercise. Fertile eggs can be better assured when the hens are compelled to scratch for every grain they get. I believe in good mashes in the morning; but at noon and night they should have grain thrown among a lot of litter — chaff, straw, leaves, or anything light — which will at once put the hens at work, not only digesting their food, but warming up the bodies. What prettier sight is there to behold than a busy hen on a cold day, with her cheery song and bright red comb? So much for the breeding stock; the work of producing the broilers, the care of the incubators and brooders, the feed, and all that pertains to the business, I have already given as plainly and ably as I knew how.

For the best results with ducks, the breeders should, like the pullets, be selected from the April and May hatches. In making the selection do not refuse a good duck simply because it is not large, for very often the best ducks are small in size. The Pekin variety is undoubtedly the best of the pure breds for market purposes, but I have more faith in the Aylesbury-Pekin cross. It would give hardier stock, and as the Aylesbury has more meat on its carcass than the Pekin (the latter, however, being the largest in size), a combination would be excellent. If it is profitable to cross breeds of poultry, why not water fowls? See that the birds have good warm houses in winter, for the more comfortable they are kept in winter, the earlier they will lay in spring. It is not necessary to have bathing water for the ducks, although where it can be had the birds will keep themselves cleaner. As soon as possible remove them to their winter quarters, so that they can settle down and "feel at home," for this constant changing quarters will prevent their early laying. Always go easily among the ducks, and do not give them any cause for alarm, for they are very shy, and very easily startled. I feed my breeding ducks two quarts bran, two quarts mixed feed, and one quart beef scraps, with about two-thirds the quantity of green stuff. My mixed feed is composed of equal parts of ground oats, corn meal, and middlings. Have regular hours for feeding, and it is wonderful how the birds will know the exact hour, and how at that time they will begin their call. My man says he knows the feeding time as well by those ducks as if he carried a watch.

There is one evil that must be guarded against in raising young ducks, and that is in not giving them sufficient room. Do not overcrowd. We lost a great many ducklings from this trouble until we got our eye teeth cut. Give them all the house and yard room they need. Mr. Hallock, on the Atlantic Duck Farm, Speonk, L. I., has his breeding pens measure 13 x 13 feet, and the runs 26 x 125 feet, of which 26 x 36 feet is water. In each of these pens are about thirty-five ducks. E. S. Grant, of Hammonton, N. J., keeps twenty-five breeders in each pen, and has house room for such a flock of 6 x 16 feet. I keep twenty breeders in a flock, and give them floor space of 8 x 10, with a run of 10 x 100 feet. I bed with salt hay; I have no board floor, but bed heavily.

NOTES IN PASSING.

George G. Harley, manager of Long View Poultry Farm, Hyattsville, Md., says that he believes his success with fowls and chicks is due to three things: tepid water, citrate of iron, and charcoal. During cold weather he takes the chill off the cold water before giving to the fowls or chicks; once a week he puts a teaspoonful of citrate of iron in the drinking water; and he always keeps charcoal before them, so they can help themselves at will.

If marketed when not more than ten weeks of age, the Light Brahma makes an excellent broiler. At eight weeks of age it will reach one and a half pounds. After ten weeks old it passes out of that plumpness, and grows more to bone and muscle, which unfits it for broiler purposes. When being dressed for market the feathers on the legs should be carefully shaved off.

Mrs. Dr. Richards, Natick, Mass., gives these weights for eleven cockerels and eleven pullets she raised up to sixty-one days old, the age at which they were weighed:

Cockerels. — Three pounds one ounce; two pounds fifteen ounces; two pounds thirteen ounces; two pounds twelve ounces; two pounds eleven ounces; two pounds ten ounces; two pounds nine ounces; two pounds nine ounces; two pounds seven ounces; two pounds seven ounces; two pounds four ounces.

Pullets. — Two pounds eight ounces; two pounds seven ounces; two pounds seven ounces; two pounds five ounces; two pounds three ounces; two pounds three ounces; two pounds three ounces; two pounds two ounces; two pounds one ounce; two pounds one ounce; two pounds.

As broilers sell in market at a pound and a half each, and as the above weights were secured in a little less than nine weeks, the evidence is very strong for the Brahma as a broiler; but the main objection to the breed is the difficulty to get good average results in hatching the eggs artificially, on account of the thickness of the shell, and the poor fertility that is general owing to the size of the fowls. Being very easily overfattened, is one of the reasons for poor hatches with Brahma eggs.

Which part of the egg makes the chicken — the white or yellow? Neither the white nor the yellow makes the chick, and yet, to a certain extent, both do. That is, the real life is nourished by both. The chick, as an individual, is neither yolk nor albumen. The germ which makes the chicken — the life and individual — is imparted by the male, and seen in the little cell located on the side of the yolk. The egg is what nourishes and grows the germ into the visible chicken that hatches. Without this germ the egg is of no account whatever, only a reservoir of nourishment that causes the growth of the germ — but one may take the ground that the egg is the chick, the male only quickening the same into life. The yolk is the last to be absorbed by the germ.

Chicks die in the shell when the hens are too fat, stock inbred, stock underfed, breeding stock diseased, eggs too old, eggs are chilled, air cells in eggs small, air cells in eggs too large, moisture too much, ventilation insufficient, temperature too low, temperature too high, air in incubator impure, egg chamber too dry when hatching, air in the room impure, and too much dampness in the cellar.

J. L. Campbell, West Elizabeth, Pa., who is one of the best authorities on incubation and the rearing of chicks, gives these hints on feed and care:

"First, never overheat them. Second, never let them get chilled; either will cause bowel troubles. It is of fully equal importance to never overfeed, or never starve; but underfed chicks will thrive better than those which are overfed.

"Do not crowd. One hundred chicks can be raised much easier in two flocks than in one; they can be raised still easier in four than in two; but one hundred chicks in a flock, if all of one age, and in a good brooder, can be raised successfully.

"The proper time to begin feeding chicks is with the hens which are to lay the eggs, say about three months before you want to set the eggs. See that the hens are in strong and vigorous health, and have got fairly started to laying. The chicks from such hens will be much easier to feed properly than from any other sort.

"I have tried everything in the shape of poultry food that I ever heard of to feed chicks with, and have succeeded in raising them on almost anything you could mention; but there is one article that I have never been able to get along without, and make what I consider a success with a brooder full of chicks, and that is fresh milk. If it is whole milk all the better, but any kind that is not spoiled is good. Sour milk is not spoiled milk; fresh buttermilk is excellent both as a drink and a food.

"The food which has always given me the best results is made from perfectly sound and sweet corn and wheat, equal proportions, coarsely ground and well mixed together.

"This we prepare in various ways, but the best possible plan is to mix it up with sour milk, sweetened with soda (a level teaspoonful to a quart of milk). Add a little New Orleans molasses or brown sugar, and as often as possible some fresh ground meat. The trimmings from the butcher shop, ground up, bone and all, are excellent. Also, a few onions cut fine, or simply sliced; half as much salt as soda, rather less than more. Mix all well together, and bake it in a hot oven from two to three hours, owing to thickness. Bake in a tightly covered pan. When you are done you will want to eat it yourself. I always have to hide it from my children, or they will eat more than the chickens, which proves that it is good, as a child knows a good article of food as well as anyone. We feed this until the chicks are at least a month old — the longer the better — and we want enough sweet milk on it to make it damp and crumbly. The same food is equally good steam cooked, nearly as dry. Any large, tightly covered vessel will answer for a steamer. Set the meal inside of it in another vessel, so the water will not boil up and mix with it.

"We begin feeding when the chicks are twenty-four to thirty-six hours old, and the above food is as good to start with as anything else. A little fine charcoal made from burnt grain or wood, is mixed with the food occasionally, and the chicks are fed raw onions several times a week. They have all the oyster shell and fine grit they will eat, which is not a little. In warm weather they have fresh water all the time, and in cold weather a little more milk and water in the food, and none to drink. If you begin watering you must never stop it.

"When a month old we begin feeding a little whole wheat, just a few grains at first; after a while this is fed at least once a day. It is excellent boiled or raw, but will go further boiled, and the chicks will thrive better on it.

"When chicks run out with the hen in warm weather, over a good large range, it makes little difference what they are fed on. They will thrive, anyway, because they get plenty of exercise and insects, which the hen finds for them; but there are two articles of food which will kill any chicks if fed improperly. One is finely ground corn meal, and the other cracked wheat. Either can be fed in small quantities, but either will pack in the crops and kill the chicks if they are given too much. It will do so for me. I have known thousands to be killed by feeding on cracked wheat, although it is an excellent food if rightly fed — which is to never let the chicks fill their crops with it.

"My plan has always been to feed three times daily, as much as they would eat up clean. I do not claim it to be better than oftener, but it is less trouble, and raises the chicks all right. The chicks, when at all confined, are given something to amuse themselves with between feeds, such as cabbage, onions, and a few boiled potatoes to pick at, and some turnips, rutabagas when we can get them. The rutabaga is an excellent food for either chicks or fowls, and can always be had in winter and spring. They have these to pick at to keep them out of mischief, such as picking at each other. Chicks are of a stirring nature, and want to be on the move all day long. The exercise counts as much as the food. When they begin to stand around half the time, the flock will be smaller very soon.

"The dry food method does well when chicks are in very small lots, and run with hens. It consists in feeding such things as dry oat meal, cracked corn, cracked wheat, millet, etc.; but I have never yet got beside any person who practiced that method of feeding brooder chicks but I beat in the end.

"No matter what we feed, we make it a point to always have it clean and pure. Never feed any spoiled food of any kind, especially spoiled meat. Always have the drinking water clean and fresh. A very little black pepper occasionally is good; the red is entirely too strong for young chicks. No matter what sort of food I have, I should never expect to make a complete success of raising brooder chicks unless I had sweet milk. I have paid ten cents a quart for it, and found that it paid me to buy it rather than to do without it.

"The smaller chicks must always be sorted out when the flock does not grow evenly. If that is not done, they will soon be trampled to death, or so stunted that it will not pay to raise them. Each lot should be kept as near a size as possible while growing.

"To get rid of gapes, skip one year entirely. Every chick we used to hatch out got the gapes. It is now sixteen years since we had a case in our yards, and all we did to eradicate them was to skip one season entirely. I did not have a chick on the place until September; then we hatched some to see if they would take them, but they did not. We refer to the gape worm, of course. The other sort is simply a cold caused by dampness. These worms remain in the ground all winter, but cannot live through the summer without chicks to breed in.

"We also make it a point to keep our small chicks inside until the dew is off the grass. After they are a month or more old they get out as soon as they like, whenever the weather is fine.

"I find that there is a very close connection between a healthy chick and the size of the air cell when being incubated. For instance, if we give the very greatest possible amount of moisture that will produce live chicks, I find that few, if any, of them will live afterwards. If we go to the other extreme, and dry the eggs down to the utmost limit which will admit of hatching the chicks, they are small and thin — never do any good, and most all will die inside of two weeks, no matter how great the care used in caring for them. But if we go about half way between these two extremes, the chicks will prove to be nice, plump and perfect: and if the stock that laid the eggs were healthy, we can raise every one, barring accidents. Now it stands to reason that to get a healthy hen we must have a vigorous and healthy chick to start with. I never yet succeeded in raising healthy, vigorous birds from puny, sickly chicks. Neither could I ever get strong, healthy chicks from unhealthy laying stock. Therefore, it can be readily seen how important it is to start right.

"Eggs which are perfectly fresh and kept at a uniform heat of 102 to 103 degrees, in ordinary warm weather, require no moisture at all. In fact, they hatch best without it, giving just a little at hatching time to prevent the chicks sticking. Then, again, those very same eggs under certain conditions of the atmosphere, would require moisture to make them hatch well.

"There is one point about the matter, however, which is very generally understood, and that is that the eggs themselves never require any moisture, hence sprinkling is nearly always injurious. It would only be beneficial when the eggs were dried down too much. It is the air which needs the moisture, simply to prevent robbing the eggs. So when the air is humid enough, no moisture whatever is needed. Now there is only one method of telling to a certainty when the eggs are right, and that is by the size of the air cells, and that point can only be determined satisfactorily by a long series of tests."

www.ingramcontent.com/pod-product-compliance
Lightning Source LLC
Chambersburg PA
CBHW082359220526
45470CB00008B/2797